Taiwo Osun

Agricultural Technology, Market Access and Welfare of Rural Households

Taiwo Osun

Agricultural Technology, Market Access and Welfare of Rural Households

Südwestdeutscher Verlag für Hochschulschriften

Impressum / Imprint

Bibliografische Information der Deutschen Nationalbibliothek: Die Deutsche Nationalbibliothek verzeichnet diese Publikation in der Deutschen Nationalbibliografie; detaillierte bibliografische Daten sind im Internet über http://dnb.d-nb.de abrufbar.

Alle in diesem Buch genannten Marken und Produktnamen unterliegen warenzeichen-, marken- oder patentrechtlichem Schutz bzw. sind Warenzeichen oder eingetragene Warenzeichen der jeweiligen Inhaber. Die Wiedergabe von Marken, Produktnamen, Gebrauchsnamen, Handelsnamen, Warenbezeichnungen u.s.w. in diesem Werk berechtigt auch ohne besondere Kennzeichnung nicht zu der Annahme, dass solche Namen im Sinne der Warenzeichen- und Markenschutzgesetzgebung als frei zu betrachten wären und daher von jedermann benutzt werden dürften.

Bibliographic information published by the Deutsche Nationalbibliothek: The Deutsche Nationalbibliothek lists this publication in the Deutsche Nationalbibliografie; detailed bibliographic data are available in the Internet at http://dnb.d-nb.de.

Any brand names and product names mentioned in this book are subject to trademark, brand or patent protection and are trademarks or registered trademarks of their respective holders. The use of brand names, product names, common names, trade names, product descriptions etc. even without a particular marking in this work is in no way to be construed to mean that such names may be regarded as unrestricted in respect of trademark and brand protection legislation and could thus be used by anyone.

Coverbild / Cover image: www.ingimage.com

Verlag / Publisher:
Südwestdeutscher Verlag für Hochschulschriften
ist ein Imprint der / is a trademark of
OmniScriptum GmbH & Co. KG
Heinrich-Böcking-Str. 6-8, 66121 Saarbrücken, Deutschland / Germany
Email: info@svh-verlag.de

Herstellung: siehe letzte Seite /
Printed at: see last page
ISBN: 978-3-8381-5074-1

Zugl. / Approved by: Doktorarbeit, Faculty of Agriculture and Nutrition Sciences, University of Kiel. January 28, 2015

Copyright © 2015 OmniScriptum GmbH & Co. KG
Alle Rechte vorbehalten. / All rights reserved. Saarbrücken 2015

Executive Summary

Several empirical studies have shown the importance of agricultural innovations on agricultural growth and poverty alleviation amongst rural households in the developing countries of the world. However, agricultural growth cannot be achieved without the adoption of productivity enhancing technologies by farmers. Despite increasing efforts by government and the international organizations to promote technology adoption in the Sub-Saharan Africa (SSA), adoption rates and application level of modern inputs have been very low and their proliferation has been slow and incomplete, hence agricultural productivity in this region remains stagnant (Matsumoto and Sserunkuuma, 2013).

In the cases where productivity enhancing technologies are adopted, not much is known about their impact on economic returns and welfare of the rural households (Minten and Barret, 2008; Omilola, 2009). While promoting adoption of productivity enhancing technologies amongst smallholder producers in the developing countries, facilitating their access to rewarding markets is very crucial in translating production into income. However, not much is known about determinants of market participation and impact of market participation on welfare amongst smallholder farmers (Bellemare and Barrett 2006; Barrett, 2008; Abdulai and Birachi, 2009).

Using a data set collected from 380 rice producing households in Nigeria, the present study investigates adoption and diffusion of agricultural technology and its impact on farm income and welfare of farm households. Specifically, the study analyzes the adoption and diffusion of New Rice for Africa (NERICA) in a dynamic framework using optimal adoption time model and duration analyses which are superior to static models widely employed in the literature. The study also examines determinants of market participation and its impact on economic returns and prosperity of farm households. Impact of technology adoption on outcomes of interest as well as determinants and impact of market participation on economic returns and prosperity of farm households are analyzed using the Endogenous Switching Regression (ESR) approach initially proposed by Lee (1982) and modified by Lokshin and Sajaia (2004). Lee's model permits estimation of the selection and outcome equations in a two-stage procedure, an approach that generates heteroskedastic residuals that cannot be used to derive consistent standard errors without cumbersome adjustments (Maddala, 1986).

A more efficient and consistent way to estimate the ESR model is the full information maximum likelihood (FIML) method (Lokshin and Sajaia 2004). Given that farm households self-selected themselves into technology adoption and market participation, ESR accounts for self-selection bias that may arise from the differences in their observable and unobservable characteristics by treating selectivity as an omitted variable problem (Heckman, 1979). The model also provides information on the differential impact of the explanatory variables on the outcomes of interest for treatment and control groups.

The empirical results on adoption and diffusion of NERICA technology show evidence of rank, stock, order and epidemic effects. Rank effect implies that potential adopters of the new technology have different inherent characteristics (such as firm size) and as a result obtain different (gross) returns from the use of the new technology. These different returns then generate different preferred adoption dates. Order effect suggests that there is existence of adoption orders which might be as a result of varying levels of access to information, production resources and special skills. Stock and epidemic effects indicate that contact with farmers who have successfully adopted the technology has been a viable medium of communicating the technology to potential adopters. Therefore, efforts to disseminate NERICA technology should take heterogeneity of farmers into consideration, and as such, tailor-made programs which best suit potential adopters with different ranks and orders should be introduced. Such programs could serve as good complements to the conventional extension system whose primary aim is to train farmers on existing technologies rather than transfer of a new technology. Interaction among farmers and social learning can be encouraged by farmers-field days and farmer-field schools. Access to credit and output markets by smallholder farmers should be facilitated as these variables can speed up rates of adoption and diffusion of NERICA technology amongst rice producers in Nigeria.

The empirical results on impact of technology adoption show that formal education, farm size and access to extension services play significant roles in technology adoption, as well as on net returns and poverty reduction amongst farm households. The results also confirm the importance of learning from social networks and access to production inputs, such as land and credit among other factors in adoption decisions and consequently on welfare of rice producing households. Adoption of NERICA technology increased net-returns by about 7.5% and reduced poverty headcount by about 35% suggesting that adoption of NERICA technology contributes significantly to farm income and welfare of rice producing households in Nigeria.

The empirical results on determinants of market participation and its impact on Return on Investment (ROI) and poverty incidence amongst farm households show that price and non-price factors such as labour, land ownership, access to credit and off-farm income, gender of household head and locational characteristics have positive and significant effects in determining market participation. Market information variables such as ownership of mobile phone and extension services also have positive and statistically significant impact on market participation. The results of the causal effect of market participation show that it increased ROI by about 33.47% and reduced poverty by about

16.46% suggesting that market participation contributes significantly to economic returns and poverty reduction amongst rice producing households in Nigeria. Market participation can be promoted through infrastructural development, especially with the view of reducing transaction costs, provisions of market information and strengthening of farmers' networks.

Table of contents

Executive Summary I
Table of contents V
List of tables VII
List of figures X
Acronyms XI
Chapter One 1
    Introduction 1
    1.1   Problem setting and motivation 1
    1.2   Study objectives 6
    1.3   Significance of the study 6
Chapter Two 11
Rice Production and Marketing in Nigeria 11
    Introduction 11
    2.1   Some background and country's profile 11
    2.2   Overview of Nigeria's agricultural sector 13
    2.3   A brief review of the agricultural policies and program in Nigeria 15
    2.4   Rice production and consumption in Nigeria 18
    2.5   NERICA dissemination and adoption in Nigeria. 23
    2.6   Marketing of domestic rice in Nigeria 24
Chapter Three 27
Literature Review 27
    Introduction 27
    3.1   Paradigms of agriculture development 27
    3.2   The uptake of agricultural innovation 29
        3.2.1   Adoption of agricultural technology 30
        3.2.2   Diffusion of agricultural technology 36
    3.3.   Impact evaluation 44
    3.4   Market participation by smallholder farmers 57
Chapter Four 61
Conceptual Frameworks and Empirical Models 61

Introduction: 61
4.1 Conceptual framework for adoption and diffusion of NERICA technology 61
    4.1.1 Duration analysis 63
    4.1.2 The empirical models for adoption and diffusion of NERICA technology 67
4.2 Conceptual framework for determinants and impact of technology adoption 69
    4.2.1 Empirical models for determinants and impact of technology adoption 70
4.3 Conceptual framework for smallholder market participation 74
4.3.1 Empirical models for determinants and impact of market participation 76
Chapter Five 79
Household Survey and Data Collection 79
Introduction: 79
    5.1 The Study Area 79
    5.2 Sampling procedure and data collection 81
    5.3 An overview of the descriptive statistics of the farm households surveyed 82
    5.4 Constraints to rice production and marketing 85
Chapter Six 87
Empirical Results 87
Introduction: 87
    6.1 Adoption and Diffusion of NERICA Technology 87
        6.1.1 Variables included in the model 88
        6.1.2 Empirical results 98
        6.1.3 Concluding remarks 105
    6.2 Impact evaluation 106
        6.2.1 Summary statistics and definition of the variables included in the model 107
        6.2.2 Determinants of NERICA adoption 110
        6.2.3 Impact of NERICA adoption on net-returns 112
        6.2.4 Impact of NERICA poverty head-count 115

    6.2.5   Impact of NERICA on Poverty Gap 118

    6.2.6   Concluding remarks 121

  6.3   Market participation... 122

    6.3.1   Summary statistics and definition of the variables included in the model 123

    6.3.2   Determinants of market participation.. 126

    6.3.3   Impact of market participation on return on investment (ROI) 127

    6.3.4   Impact of market participation on poverty head-count 130

    6.3.5   Impact of market participation on poverty gap 132

    6.3.6   Concluding remarks 135

Chapter Seven... 137

Summary and conclusion 137

Introduction: 137

  7.1   Summary of findings 137

  7.2   Policy implications of the study 138

References 140

List of Tables

Table 2-1	Summary of rice production systems in Nigeria	20
Table 5-1	Descriptive statistics of farm households surveyed	83
Table 6-1	Ordinary least square estimates of the parameter of profit function	90
Table 6-2	Maximum likelihood estimates of determinants of access to credit	94
Table 6-3	Maximum likelihood estimates of determinants of access to extension	95
Table 6-4	Descriptive statistics and definition of the variables used in adoption and diffusion model	97
Table 6-5	Maximum likelihood estimates of the parameters of hazard models	103
Table 6-6	Maximum likelihood estimates of the parameters of hazard ratio	104
Table 6-7	Summary statistics of farm and household characteristics of adopters and non-adopters of NERICA technology	109
Table 6-8a	Endogenous Switching Regression results for determinants of NERICA adoption and impact on net-returns	114
Table 6-8b	Impact of NERICA adoption on net-returns (ATT)	115
Table 6-9a	Endogenous Switching Regression results for determinants of NERICA adoption and impact on poverty-headcount	117
Table 6-9b	Impact of NERICA adoption on poverty-headcount (ATT)	118
Table 6-10a	Endogenous Switching Regression results for determinants of NERICA adoption and impact on poverty-gap	119
Table 6-10b	Impact of NERICA adoption on poverty-gap (ATT)	120

Table 6-11	Summary statistics of farm and household characteristics of market participants and non-participants	125
Table 6-12a	Endogenous Switching Regression results for determinants of market participation and impact on ROI	128
Table 6-12b	Impact of market participation on ROI (ATT)...................	129
Table 6-13a	Endogenous Switching Regression results for determinants of market participation and impact on poverty-headcount	131
Table 6-13b	Impact of market participation on poverty-headcount (ATT)	132
Table 6-14a	Endogenous Switching Regression results for determinants of market participation and impact on poverty-gap	133
Table 6-14b	Impact of market participation on poverty-gap (ATT)	134

List of figures

Figure 2-1:	Map of Nigeria showing population density	12
Figure 2-2:	Sectorial contribution to Nigeria's GDP	14
Figure 2-3:	Rice production and yield in Nigeria	21
Figure 2-4:	Milled rice importation into Nigeria	22
Figure 2-5:	Rice marketing channels in Nigeria	25
Figure 3-1:	Categorization of innovation adopters	39
Figure 3-2:	Logistic or sigmoid diffusion curve	40
Figure 5-1:	Map of Nigeria showing the study area	80
Figure 5-2:	Distribution of area of land under rice cultivation by farm households	84
Figure 5-3:	Constraints to rice production in Nigeria	86
Figure 6-1:	Adoption pattern of NERICA technology	88
Figure 6-2:	Estimate of survival function for the whole sample	99
Figure 6-3:	Effects of FBG on survival function	99
Figure 6-4:	Effects of interaction with peers and neighbour on survival function	99
Figure 6-5:	Log rant test for equality of survival function	99

Acronyms

ADP	Agricultural Development Program (Nigeria)
AMA	American Marketing Association
ARC	Africa Rice Centre
BOI	Bank of Industry (Nigeria)
CBN	Central Bank of Nigeria
DIFFRI	Directorate of Food Roads and Rural Infrastructure (Nigeria)
FAOSTAT	Food and Agriculture Organization Corporate Statistical Database (FAOSTAT)
FDG	Focus Group Discussion
FGT	Foster-Greer-Thorbecke Poverty Measure
GDP	Gross Domestic Product
IFDC	International Fertilizer Development Centre (IFDC)
IFPRI	International Food Policy Research Institute
IMF	International Monetary Fund
MDG	Millennium Development Goals
NACB	Nigerian Agricultural and Cooperative Bank
NACRBD	Nigerian Agricultural and Rural Development Bank
NAFPP	National Accelerated Food Production Programme (Nigeria)
NARES	National Agricultural Research and Extension System (Nigeria)
NBS	Nigerian Bureau of Statistics (Nigeria)
NEEDS	National Economic Empowerment and Development Strategy
NERICA	New Rice for Africa

NRDS	National Rice Development Strategy (Nigeria)
OFN	Operation Feed the Nation (Nigeria)
PVS	Participatory Varietal Selection
ROI	Return on Investment
SAP	Structural Adjustment Program
SSA	Sub-Saharan Africa
UN-ECA	United Nations Economic Commission for Africa
UNIDO	United Nations Industrial Development Organization
WD	World Bank
WTO	World Trade Organization

Chapter One

Introduction

1.1 Problem setting and motivation

Poverty remains a major development problem in many developing countries of the world. Although tremendous progress has been made with respect to achieving the United Nations' Millennium Development Goal of reducing the number of poor people at the start of the Millennium to half by 2015, there are clear indications that the goal may not be achieved in many low and middle income countries, as the number of poor and hungry people in these countries is still significantly high. Available statistics show that, even though poverty incidence in the sub-Saharan Africa (SSA) drastically reduced by about 9.41 percent between 1999 and 2012 (from 57.89% to 48.48% of the population), yet almost half of the population in the geographic region remains poor.

In Nigeria, about 62 percent of the country's population lives on less than $1.25 a day (World Bank, 2013). Poverty is endemic in rural Nigeria, where the inhabitants depend on agriculture for survival and have limited access to social services and infrastructure (IFPRI, 2010). The situation can be described as a vicious cycle of low productivity, low income and lack of purchasing power to acquire basic necessities of life.

The World Bank (2005) described poverty as inability to attain minimal standards of living measured in terms of meeting the basic human needs or possession of income required to satisfy them. The basic human needs include food, safe drinking water, sanitation facilities, health, shelter, education and information. However, the role of agricultural growth in poverty reduction in the developing countries has been widely documented in literature. The Norman

Borlaug's initiative, which would latter translate into Green Revolution was primarily designed to develop disease resistant and high-yielding wheat varieties for Mexican farmers in the 1940s in response to problem of low wheat productivity. The success of the initiative led to its replication in the then other developing countries of Latin America and Asia, resulting in impressive poverty reduction and national economic development in the 1960's and 70's (Pinstrup-Andersen and Hazell, 1985).

Several empirical studies have shown the importance of agricultural growth in poverty alleviation and stimulating overall economic growth in the developing countries of the world. For instance, Ravallion and Chen, (2007) showed that much of the progress made in poverty reduction by China between 1980 and 2001 can be largely attributed to growth in the agricultural sector than either in the secondary or the tertiary sectors of the economy. In fact, the agricultural sector had a 3.5 times larger impact on poverty reduction than the other sectors of China's economy during the period.

Using a dataset spanning over 25 years from 42 developing and transitional countries of the world, Ligon and Sadoulet (2008) also demonstrated that GDP growth originating from agriculture has a much larger positive effect on expenditure gains by the poorest households than growth originating from the rest of the economy. However, de-Janvry and Elisabeth (2002) pointed out that agricultural growth can only be achieved through productivity gains, which come mostly from land and labour productivity nevertheless, productivity gains cannot occur without the adoption of productivity enhancing technologies by farmers.

Generally, productivity gains from adoption of improved agricultural technologies can have direct and indirect effects on poverty reduction. While the direct effects of adoption of improved agricultural technologies may include household food security, lower cost of production and higher economic returns from sales of farm produce; the indirect effects include increase in the real income of both rural and urban population as a result of lowered food prices, increased employment and wages for farm labour, availability of cheap raw materials required for rapid industrialization, foreign exchange earnings and overall economic growth (Haggblade et al., 1989; de-Janvry and Sadoulet, 2002).

One of the notable productivity enhancing technologies in recent time is New Rice for Africa (NERICA). NERICA varieties are crossbreeds of African rice, "*Oryza Glaberrima*" and Asian rice, "*Oryza Sativa*" resulting in progenies with yield potentials that are three times higher than the conventional African rice species. NERICA was created by the Africa Rice Centre with the intention to solve the problems of low productivity and the continuous short fall of rice production in Africa (Africa Rice Centre, 2008). This breakthrough won Monty Jones, the lead breeder of NERICA the 2004 edition of the World Food Prize.

While NERICA technology has been relatively adopted in other African countries, smallholder farmers in Nigeria are slow in switching from the indigenous rice varieties (notably *Ofada and Igbemo* in the Southwestern part of the country) to the new and improved varieties (Bzugu et al. 2010; Dontsop-Nguezet et al 2013). For instance, the NERICA varieties are presently being cultivated in many parts of Uganda. In fact, rice acreage increased by six fold within six years - from 6,000 hectares in 2002 to 40,000 hectares in 2008, and the number of rice growers rose from 4,000 in 2004 to 35,000 in 2007, following the release of the NERICA varieties. Consequently, rice importation reduced drastically from 60,000 tons in 2005 to 35,000 tons in 2007, an almost 50 percent reduction, leading to savings of about $30 million (Akintayo et al.

2009). In the same vein, about 40% of the rice farmers in the Gambia have successfully adopted NERICA technology (Dibba et al. 2012). However, NERICA adoption rate in Nigeria is found to be only 19% (Nguezet et al. 2013).

Rice is an important staple whose popularity and consumption have been on a steady increase in Nigeria in the last four decades. Its consumption has risen tremendously as a result of the accelerating population growth, rapid urbanisation and changing family and occupational structure (IFDC, 2008). The staple is the fourth most important crop in terms of calorie consumed in the country, following sorghum, millet and cassava. However, larger proportion of the rice consumed in Nigeria is imported. Currently, the country is the second largest importer of rice in the whole world after the Philippines (Cadoni and Angelucci, 2013). Given the availability of suitable land and climatic condition for rice production, a major setback to adequate production of rice in Nigeria is low productivity. Rice yield in Nigeria is 1.80 tons/ha, compared to 5.6 tons/ha in Vietnam and 4.38 tons/ha in Bangladesh (FAOSTAT, 2013). There is therefore a need for studies to understand the adoption behavior of rice farmers in Nigeria in order to recommend appropriate agricultural policies for speedy adoption of NERICA technology.

Since poverty in Nigeria has been described as rural phenomenon arising from low productivity and income, the present study examines adoption and diffusion of New Rice for Africa (NERICA) in Nigeria, and its impact on economic returns and ultimately on poverty alleviation amongst the rural households, given that the technology offers opportunities for increased productivity, higher producer income and food security at household level.

Another major limiting constraint being faced by smallholder farmers in the developing countries is access to rewarding markets (World Bank 2002;

Dorward et al., 2005). For example, a typical smallholder arable crop farmer in Nigeria does not produce for an identified market but rather, anticipates that when his crops are mature, he would find markets for them. This is because the environment within which smallholder producers operate is characterized by many constraints which make market participation increasingly difficult for them. While constraints to efficient marketing systems in the developing countries can be classified into institutional, infrastructural, socioeconomic or economic factors (Kydd and Dorward, 2004), market participation by smallholder farmers is mainly constrained by high transaction costs and missing markets (Omamo, 1998; Ouma et al. 2010).

Although empirical studies on commercialization of smallholder producers in Africa is just building up in the literature, studies have shown that improved market access can go a long way in enhancing competitive production and producer prices (von Oppen, et al. 1997; Romer 1994). However, not much is known about determinants and impact of smallholder market participation on economic returns and welfare of the rural households. Thus, the present study also investigates determinants of market participation and its impact on Return on Investment (ROI) and prosperity of farming households in Nigeria. In particular, the roles of transaction costs and market information in market entry and production of marketed surplus are examined, while policy options for market development are suggested.

1.2 Study objectives

The general objective of this study is to examine adoption and diffusion of agricultural innovations. The study is also designed to provide explicit information on impact of technology adoption on welfare of farm households as well as smallholder market participation and its impact. Specifically, the present study

1. analyzes farm and non-farm factors influencing adoption and diffusion of New Rice for Africa (NERICA) in Nigeria.
2. examines impact of adoption of NERICA technology on economic returns and welfare of rice farming households in Nigeria.
3. investigates determinants and impact of market participation on economic returns and prosperity of rice producing households in Nigeria.

1.3 Significance of the study

Adoption and diffusion of agricultural innovations are essential to technological change, food security and poverty reduction in the developing countries. Although empirical studies on determinants of adoption behavior have received considerable attention in the literature, studies on diffusion of agricultural technologies are scanty. Similarly, it has been argued that empirical evidence of impact of agricultural technology adoption on welfare of rural households in Africa is not convincing (Minten and Barret, 2008; Omilola, 2009). Although NERICA technology offers opportunities for increased productivity and food security, not much is known about its farm level productivity and impact on welfare of farm households. In the same manner, market participation behavior amongst smallholder producers has not received adequate attention in the literature as not much is known about determinants of market participation and its impact on welfare of farm households.

The present study is therefore unique and contributes to knowledge in three folds. First, it examines adoption and diffusion of agricultural technology in a dynamic framework using optimal adoption time models and survival analysis which are superior to static models widely employed in the literature. A general class of models known as proportional-hazard models proposed by Cox (1972) having the advantage of ensuring a positive hazard rate without imposing further restrictions on the parameters of the model is employed in the study. More so, time waited before adoption is examined using discrete time duration model, which is a significant improvement over adoption and diffusion studies in which time to adoption is considered within continuous time specification (see Abdulai and Huffman, 2005; Genius et al. 2014).

In particular, the study investigates how farmers' adoption decisions respond to the actions of other farmers in their information networks which appear to be ambiguous in the literature. For instance, Conley and Udry (2010) argued that potential adopters of a new technology are likely to adopt the technology after learning about its characteristics from their neighbors, while Munshi (2004) argued that although a potential adopter may probably have watched his neighbors successfully adopting a new technology, he may choose not to adopt the technology especially when the characteristics of the pioneer farmer is different from his. To this extent, the present study provides new insights on the roles of learning from social networks on technology adoption.

Second, it makes a significant contribution to empirical findings on impact of technology adoption on economic returns and welfare of smallholder farm households in the developing countries by using an appropriate econometric procedure. The true impact of a program or treatment (e.g., adoption of agricultural technology) is the difference between outcome (e.g productivity) due to exposure to treatment and the counterfactual situation. The counterfactual situation is the outcome that would have resulted in the absence of treatment. In randomized experiment, assignment into treatment and control group is random, so the control group has the same distributions of both observed and unobserved characteristics as the treatment group and as such, the control group provides a suitable counterfactual (Rossi and Freeman, 1993).

However, when the data available for a study are from a cross-sectional survey, such as the one employed in the present study, there would not be information on the counterfactual situation because it is practically impossible to observe outcomes for an individual in the two states (i.e., factual and counterfactual situations). At the same time, we cannot simply use a non-treatment group as control group (or counterfactual) due to self-selection problems. The present study therefore employs the Endogenous Switching Regression (ESR) approach to account for self-selection that may arise from observable and unobservable characteristics of the rice producers, in order to consistently estimate impact of technology adoption on outcomes of interest (Lee, 1982; Lokshin and Sajaia 2004). ESR is a generalization of Heckman's model (1979), in which sample selection is treated as a problem of specification error or omitted variable, which can be corrected by explicitly using information gained from the selection equation for consistent estimation of the outcome equation (Shenyang and Fraser 2010).

The major advantages of ESR is that information is provided on determinants of technology adoption, the differential impact of the explanatory variables on outcomes of interest for adopters and non-adopters as well as treatment effects of adoption. Finally, rice market participation by smallholder farmers has not been accorded adequate attention. The present study therefore explicitly investigates determinants and impact of market participation on economic returns and welfare of rice producing households in Nigeria.

1.4 Outline of the thesis

The dissertation is organized as follows;

Chapter one gives general introduction, problem setting and motivation of the study, as well as the study objectives.

Chapter two provides a brief background of Nigerian agricultural sector and an extensive overview of rice production and marketing activities by smallholder farmers in the country.

In chapter three, a comprehensive review of the literature on the three thematic areas of the study; (1) adoption and diffusion of agricultural innovation (2) impact of adoption of agricultural technology (3) determinants and impact of market participation are presented, while the identified research gaps in the literature are highlighted at the concluding part of each sub-section.

Chapter 4 showcases the conceptual frameworks and analytical models for adoption and diffusion of agricultural technology, impact evaluation and market participation respectively. Karshenas and Stoneman (1993) optimal time framework is employed to analyze the relative importance of farm and non-farm factors (i.e., stock, rank, order and epidemic effects) influencing adoption and diffusion of NERICA technology while discrete time proportional hazard model is employed to examine timing of adoption of NERICA technology. Impact of technology adoption as well as market participation and their impact on welfare of farm households are analyzed by employing the quasi-experimental approach of endogenous switching regression method which makes it possible to account for selection bias that may arise from observable and unobservable characteristics of the farm households and to estimate the differential impact of the explanatory variables on the outcomes of interest respectively.

Chapter five provides information on data collection procedures and descriptive statistics of the data generated during field survey.

Finally, chapters six and seven present the results of the econometric estimations, conclusions and policy recommendations respectively.

Chapter Two

Rice Production and Marketing in Nigeria

Introduction

This chapter provides some background information about Nigeria, an overview of the agricultural sector as well as general information about rice production and marketing activities in Nigeria.

2.1 Some background and country's profile

Nigeria is a tropical country located between the equator and the tropic of cancer. The main latitude and longitude of Nigeria is 10°North and 8°East. The climate varies from equatorial in the South, tropical in the Central to arid in the North, while the terrains are low lands towards the South, hills and plateaus in the Central and plains in the North.

The weather varies with the rainy and dry seasons, depending on location; the length of the rainy season decreases from South to North. In the South the rainy season lasts from March to November, whereas in the far North it lasts only from mid-May to September. Precipitation is heavier in the South, especially in the Southeast, with about 120 inches (3,000 mm) of rain a year, and lowest in the Northern part of the country with about 20 inches (500 mm) a year. Temperature and humidity remain relatively constant throughout the year in the South, with a mean temperature of 30^0C, while it varies considerably with seasons in the North – for example, in the Northeastern city of Maiduguri, the mean monthly temperature is about 38^0C (Encyclopedia Britannica).

Nigeria occupies a total area of 92.38 million hectares, consisting of 91.08 million hectares land area and 1.3 million hectares inland waters. The agricultural land is about 72 million hectares; out of which 35 million hectares (48%) are arable land, 6.7 million hectares (9.3%) are under permanent crops while 8.22 million hectares are forest (11.41%) (FAOSTAT, 2013). The country shares land borders with the republics of Benin in the West, Chad and Cameroon in the East, and Niger in the North. Its coast lies on the Gulf of Guinea in the South and borders Lake Chad to the Northeast.

On 1 October 1960, Nigeria gained independence from Great Britain and now, the country is federation of 36 States, with the Federal Capital Territory (FCT) in Abuja. The 36 States are further sub-divided into 744 Local Government Areas for efficient grassroots administration. Nigeria is Africa's most populous nation, with an estimated population of 177 million people which is composed of 250 ethnic groups and an estimated population density of 173.94 people per sq. km. The major ethic groups are; Hausa and Fulani (29%), Yoruba (21%), Igbo (18%), Ijaw (10%), Kanuri (4%), Ibibio (3.5%) and Tiv (2.5%). Presently, Nigeria is Africa's largest economy, with 2013 GDP estimated at USD 502 billion. Crude oil exploration has been a dominant source of government revenue since the 1970s, while the country's economy has continued to grow at a rapid rate of 6-8% per annum (CIA, 2014).

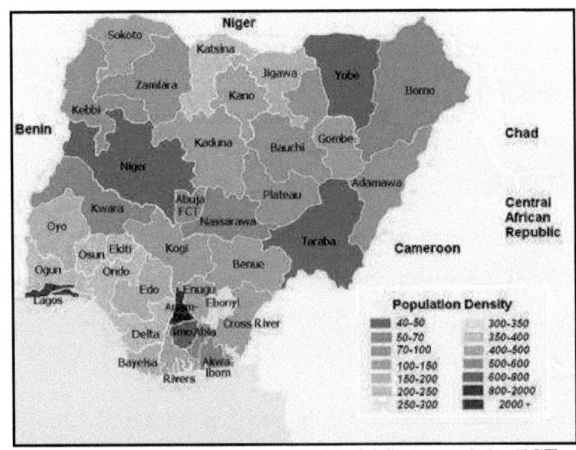

Figure 2-1: Map of Nigeria showing the 36 States and the FCT

In the periods before 1970s, the agricultural sector was the most important sector of Nigeria's economy in terms of contributions to domestic production, employment and foreign exchange earnings. The situation remained almost the same three decades later with the exception that agriculture is no longer the principal foreign exchange earner, a role now being played by the crude oil sector. However, oil wealth has not translated into evenly distribution of wealth, as about 62% of the population lives on less than $1.25 a day. In terms of employment, agriculture is still by far the most important sector of Nigeria's economy, engaging about 70% of the labor force.

Poverty is endemic in rural Nigeria where the inhabitants depend on agriculture for survival and have limited access to social services and infrastructure. The situation can be described as a vicious cycle of low productivity, low income and lack of purchasing power to acquire basic necessities of life (IFPRI, 2010; NBS, 2014). Given that agriculture still remains an important sector of Nigeria's economy, Nigeria is a signatory to "Maputo 2003 Declaration" on agricultural development and food security. The Maputo Declaration anticipates an African Green Revolution through a strong commitment to development of the agricultural sector by setting aside at least 10% of the annual national budget for agricultural and rural development activities.

2.2 Overview of Nigeria's agricultural sector

Historically, agriculture was the mainstay of Nigeria's economy and the primary foreign exchange earner. The country produced and exported large volumes of cocoa, cotton, palm oil, palm kernel, ground-nuts and rubber in the 1950s and 1960s, while government revenues were heavily dependent on taxes from those exports. However, as soon as large scale exploration of crude oil began in the 1970s, the sector started declining, while the share of agriculture in the GDP reduced from 60% in the early 1960s through about 40% in the 1970s and even lower thereafter. As of 2013, the agricultural sector contributed only 24.39% to the GDP, while industries and services sectors contributed 22.24% and 53.37% respectively, (see figure 2-2).

Similarly, agricultural exports which constituted over 80% of the country's total export in 1960s, reduced to less than 10% in 1980s and presently, oil and gas constitutes about 97% of exports and 85% of government revenues (Daramola et. al, 2007; United Nations 2013; NBC, 2014). The relative neglect of the agricultural sector can be attributed to the booming oil sector, unstable and inappropriate macroeconomic policies (of pricing, trade and exchange) and inappropriate agricultural policies (Etim and Edet, 2013).

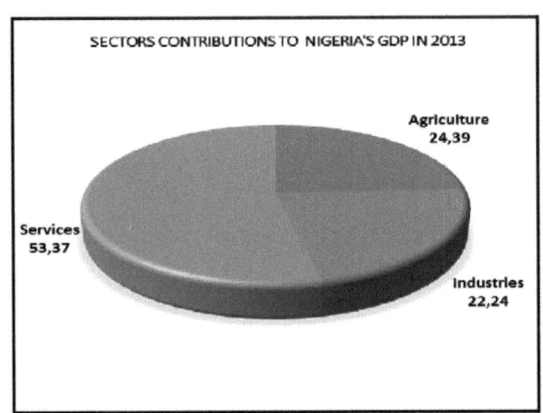

Figure 2-2: Sectorial Contributions to Nigeria's GDP in 2013
Source: Source NBS, 2014

Nigeria's agriculture is faced with a myriad of problems such as limited access to inputs and credit facilities (Oni et al., 2009), poor extension services (Igben and Nwosu 1987), low adoption rates (Adeoti and Sinh, 2009), vague agricultural policies and poor infrastructure (Okuneye, 1990). The sector is characterized by the use of simple farm tools, low productivity, poor postharvest handling, inefficient marketing systems and weak value-chain linkages

(UNIDO, 2010), pests and diseases (Oruonye and Okrikata 2010) as well as high production and transaction costs (National Rice Survey 2009).

The sector is dominated by smallholder farmers who cultivate small-scattered plots of land less than 2 hectares and depend mainly on rainfall for their production activities rather than irrigation systems. All the same, these smallholder farmers account for over 90% of food production in Nigeria. Agricultural activities are in primary production with limited value-addition, processing and packaging. Given the inherent problems in the agricultural sector, Nigeria spends huge amount of money on importation of food commodities to augment shortfall in production. Major food commodities imported include wheat, rice, sugar, malts and maize (Olomola, 2007).

2.3 A brief review of the agricultural policies and program in Nigeria

Agricultural development in Nigeria started with the establishment of a botanical garden in Lagos in 1893 with the aim of introducing exportable crops like oil palm, rubber, cotton and cocoa to farmers (Roseboom et al. 1994). Considerable emphasis was placed on research and extension of these exportable crops which were mainly sold to British trading companies. In the late 1950's, the agricultural commodities marketing boards were established in response to the monopolistic practices of the trading companies however, marketing boards soon became instrument of taxation and viable source of government's revenue.

As argued by Blandford (1979), the use of the marketing boards in most West African countries as a fiscal device was not the original purpose of their creation. For example, in the early years of the Nigerian Cocoa Marketing Board, only about 2% of the board's total sales value accrued to government as tax but by the fifth year, this had increased to 20%. As such, much emphasis were laid on the development of exportable crops sector at the expense of food production, as taxes from the marketing boards were major sources of government's revenues.

Soon after large scale crude oil exploration began, there was a sharp decline in food production due to a relative neglect of the agricultural sector and massive rural-urban migration. As a result, food production could not meet up with the growing population. The situation became worsened with the civil war which occured between 1967 and 1970 in Eastern Nigeria, resulting in famine outbreak in the region during and after the war. About the same time also occurred the Sahelian drought which seriously affected agricultural production in Northern Nigeria, where most of the grains and livestock were produced (Busch 1988).

In order to salvage the situation, the National Accelerated Food Production Program (NAFPP) was introduced in 1972 to transfer relevant technologies for accelerated production of important food crops such as rice, maize, sorghum, millet and wheat. The initiative was designed to ensure access to cheap staple foods and curtail imports. NAFPP was replaced by Operation Feed the Nation (OFN) in 1976. In addition to the technology transfer activities of NAFPP, OFN was designed to make subsidized inputs available to smallholder farmers. However, the program suffered some setbacks due to corruption of the implementing personnel (Okuneye, 1990).

The River Basin Development Program and Land Use Act were introduced simultaneously in 1978 to facilitate food crop production through increased access to irrigation facilities in order to mitigate the impact of fluctuating rainfall, while Land Use Act was enacted to solve the inherent problem of traditional land tenure system where lands are seen as communal properties and family inheritance thereby limiting land availability for agricultural purposes. The Act operated by acquiring large expanse of land for agricultural purposes.

However, both programmes failed due to managerial problems and unnecessary political interference as a result of corruption which had permeated the nation's socio-political, economic and cultural institutions (Akindele and Adebo, 2004).

A relatively successful initiative in Nigeria's agricultural sector is the Agricultural Development Program (ADP) initiated with a World Bank loan. Like the previous programs, the ADP was designed to accelerate food production in response to the problems of food shortages, fall in agricultural productivity and lack of necessary infrastructure in the rural areas. The ADP commenced gradually in 1974 in Northern Nigeria as enclave projects in Funtua, Guzau and Gombe. Successes of these experiments led to the establishment of the program in all the States of the Federation.

Upon creation of the State ADPs, the roles of agricultural extension, provision of rural infrastructure and linkage to inputs were subsequently transferred from the Ministries to the newly created State-level ADPs to avoid duplication of tasks. However, according to Chukwuemeka and Nzewi (2011), the top-down approach that excludes the beneficiaries from participating in program design and implementation limited the success of the program. They also argued that the program is more or less a replication of the agricultural development programs implemented in some Asian countries without taking due consideration of the environmental, socio-political and economic factors of the people of Nigeria.

A micro-economic policy that had a positive and significant impact on agricultural growth in Nigeria is the Structure Adjustment Program (SAP). SAP was introduced in 1986 to promote economic growth by diversifying the country's export base away from oil dominance through the promotion of locally made goods as recommended by the International Monetary Funds (IMF) and the World Bank. The policy encouraged local production and put many imported products on prohibition list, subsidies to agriculture were removed and all marketing boards were abolished. Before it was terminated in 1993, SAP had good impact on Nigeria's agricultural sector because it increased food production and curtailed imports during its implementation (Shimada, 1999).

Other notable agricultural policies and programmes implemented in Nigeria include rural banking program operated by the Nigerian Agricultural and Cooperative Bank (NACB, latter NACRBD), Green Revolution (1980 – 1983) and the Directorate of Food, Roads and Rural Infrastructure (DFRRI) (1987 – 1993). Equally, the Presidential Initiatives for accelerated production of major crops like cassava, rice, vegetable oil and tree crops (2000 – 2007), Fadama I and II, the National Economic, Empowerment and Development Strategy (NEEDS) (2004 – 2010) and lately, the agricultural transformation agenda (ATA) were introduced to solve the inherent problems of increasing food shortages and importation.

There is no doubt that the Nigerian agricultural sector has witnessed many developmental policies and programs, however despite all these programs, poverty and food insecurity still remain fundamental problems in the country. Perhaps the reasons why these problems have not been solved is corruption and political instability. However, many researchers have argued that lack of necessary data and empirical studies for proper planning might be major contributing factors.

2.4 Rice production and consumption in Nigeria

Rice is a food crop whose popularity and consumption have been on a steady increase during the last three decades in Nigeria. Its consumption has risen tremendously as a result of the accelerating population growth rate, rapid urbanisation and changes in family and occupational structures. In both urban and rural areas, rice is consumed almost on daily basis and it accounts for more than 20% of all meals consumed per week by a typical household (IFDC, 2008). Rice is the fourth most important crop in terms of calorie consumed following sorghum, millet and cassava. Presently, rice is grown on approximately 3.7 million hectares, covering about 10.6 % of the 35 million hectares of the land under agricultural production and about 5.3% of the total arable land area available in the country (Cadoni and Angelucci, 2013).

Rice can be cultivated in virtually all Nigeria's agro-ecological zones; from the mangrove and swampy ecologies of the Niger-Delta in the central coastal areas, to the dry zones of the Sahel in the North. Specifically, there are five major rice production systems in Nigeria, these are; upland, hydromorphic, lowland, deep inland water and mangrove swamp production systems. While the upland rice is grown on free-draining soils where the water table is permanently below the roots of the rice plant, the irrigated-upland rice production is practiced in places where rainfall regime is short and as such, some forms of supplementary irrigation may be required to ameliorate drought conditions during critical stages of growth.

The hydromorphic conditions occur when water is supplied to the rice crop by a shallow ground water table within the rooting zone of the plants. Hydromorphic rice is found either on lower slopes in the toposequence or in situations where impermeable soil layer reduces water percolation. Two sub-types of lowland ecologies are available; shallow fadama and deep fadama.

A distinguishing feature of the shallow system is that the soil must be covered completely by water at some stage in the growth cycle. The deep inland water rice production is floating rice system, even when rice fields become flooded, the plants send down their deep roots into the soil while the vegetative parts float on top of the water. Under this production system, rice is planted by direct seeding or transplanting of seedlings which had been raised in a nursery. The mangrove swamp rice production is carried out in the coastal swamp areas (Longtau, 2003). Table 2-1 summarises the five rice growing environments found in Nigeria.

Table 2-1: Summary of Rice Production Systems in Nigeria

Type	Characteristics	Geographic Spread
Upland	Rainfed rice is grown on free-draining fertile soils; irrigated upland system is practiced in places where rainfall regime is short.	Widespread, except in coasts, high rain forest and sahel zones.
Hydromorphic	Rainfed rice is grown on soils with shallow ground water table or an impermeable layer. This is sometimes called wet upland.	Widespread at the fringes of streams and intermediate zones between upland and swamps of rivers in the savannah.
Lowland	Rainfed or irrigated rice in aquatic conditions or medium ground water table. Water covers soil completely at some stage during the cropping season. Lowland ecology is also known as shallow swamps or fadama.	Widespread from high rain forest to sahel zones.
Deep Inland Water	Rainfed rice grown on soils with deep water tables. The rice crop float at some stage and harvesting may be done from a canoe. The ecology is also called deep fadama or floodplain.	Found in the Sokoto-Rima basin and Chad basin, floodplains of Niger, Benue, Kaduna, Gboko, Hadejia and Konadugu-Yobe.
Mangrove Swamp	Rice is grown at the coast or swamps of the high rain forest.	Coastal areas and Warri area in Delta State.

Source: Longtau, 2003.

Although Nigeria's rice sector has witnessed some remarkable development particularly in the last 10 years, domestic rice production has not increased sufficiently to meet demand. The annual domestic rice production is about 3.05 million metric tons, while demand is about 5 million leaving a huge gap of about 2.20 million metric tons to be filled by imports. Given the availability of suitable land and climatic condition for rice production, a major setback to adequate production of rice is low productivity. While production quantity continues to grow appreciably, rice yield (productivity per unit land) is rather declining (figure 2-3).

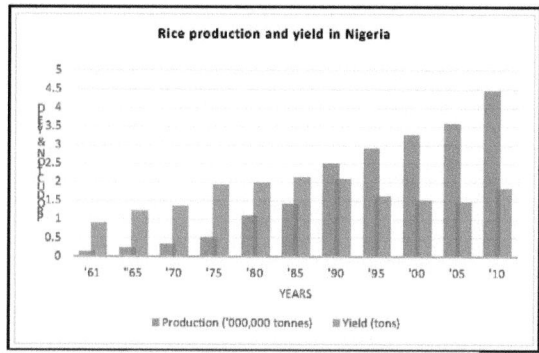

Figure 2-3: Rice production and yield in Nigeria
Source: FAOSTAT, 2013

Land expansion which had historically been the main source of growth in rice production may no longer be sufficient in the face of alarming rate of population growth. Besides, land is becoming scarce faster in almost every part of Nigeria. The problem of low productivity of rice can be attributed to low rate of adoption of high yielding varieties and low resource productivity (Nguezet et al. 2013), land degradation and poor land preparation (Kebbeh et al., 2003), unreliable and uneven distribution of rainfall (Oteng and Sant'Anna 1999), problem of pests (such as weed, insect, diseases, birds) (Nguezet et al. 2011), the use of low production technology methods and poor extension service (Longtau, 2003).

Consequently, Nigeria imports rice worth well over US$ 1 billion annually (figure 2-4).

Figure 2-4: Milled rice importation to Nigeria 1961-2011
Source: FAOSTAT, 2013.

In a bid to address the problem of shortfall in rice production, Nigerian government at various times has come up with different policies and program. However, as observed by Okoruwa and Ogundele (2004), rice production intervention policies have not been consistent. For instance, during the Structural Adjustment Program (SAP) era (1986 – 1993), rice importation was banned, the ban was lifted in 1995 in line with WTO agreement on trade liberalization. Since the ban was lifted, government has resorted to the use of tariff measures. Over the time, the tariff on rice has increased from 50% between 1996 – 1999 to 100% in 2002 and 150% in 2003. In 2009, the tariff was adjusted down to 30% for milled rice and 10% for brown rice.

Nigerian government is currently collaborating with the African Rice Centre to implement a new policy tagged National Rice Development Strategy (NRDS). NRDS is targeted at raising paddy production to 13 million tonnes in 2018. There are three priorities areas in the NRDS and these are: (I) post-harvest processing and treatment; (II) irrigation development; and (III) input availability: mainly focusing on seeds, fertilizer and farming equipment. NRDS includes a mixture of input supply promotion (such as 50% subsidy for seeds and 25% for fertilizer) and reduced custom tariff on importation of specific agricultural machineries (such as tractors and processing equipment).The National Agricultural Seed Council is in charge of seed production and certification, while the National Cereals Research Institute (NCRI) and the Africa Rice Centre regulate seeds delivery to producers (NRDS, 2009).

2.5 NERICA dissemination and adoption in Nigeria.

The New Rice for Africa (NERICA) was developed by the Africa Rice Center in 1996 with the objective of increasing productivity and reducing the continuous short fall of rice in Africa. The new rice varieties which are mostly suited to the upland rice ecologies, require no special inputs, have short growth cycle, they are pests resistant and have good vegetative growth – which make them to be highly weed competitive. Generally, the NERICA varieties are highly responsive in low input and rain-fed agricultural systems compared to other existing rice varieties in Africa (Africa Rice Center, 2008).

NERICA was introduced to Nigerian rice farmers in 2001 by the Africa Rice Center in collaboration with National Agricultural Research and Extension System (NARES) through the 'Participatory Varietal Selection' approach (PVS) and the "Training & Visit" extension system which is the official agricultural extension method in Nigeria. The Participatory varietal selection approach arose from the realization of the fact that farmers were not using crop varieties developed and tested on research stations because they think the improved germplasm may not work-out well in the real world. So farmers continued to grow old and unproductive varieties. PVS was developed in the 1980s to encourage the adoption of high yielding varieties by low-resource farmers using participatory approach. The needs of farmers are identified by discovering what crops and varieties they grow, and what traits they consider important. Scientists then select new varieties that have the traits that farmers desire and that match the farmers' landraces for important characters such as early maturity, plant height and seed type (Witcombe 1996).

PVS was implemented by establishing on-farm demonstration plots in farming communities, where traditional rice and NERICA varieties were cultivated with full participation of farmers, researchers as well as extension workers. Farmers were asked to compare the agronomic characteristics, grain quality, ease of processing and palatability of NERICA varieties to the existing rice varieties in their communities, thereafter farmers were encouraged to test some of the NERICA varieties on their fields. In 2004, NERICA dissemination received a boost through the African Rice Initiative with funding from African Development Bank and other partners for awareness creation, training of farmers and extension staff, production of certified seeds and provision of basic infrastructure such as processing mills and feeder roads in the rural areas (Ogun-State MANR, 2012; Jones et al., 2002).

2.6 Marketing of domestic rice in Nigeria

Marketing is the process of planning and executing the conception, pricing, promotion and distribution of ideas, goods and services to create exchange and satisfy individual and organizational objectives (AMA, 1985). Ihene (1996) defined rice marketing as the performance of all business activities in the flow of paddy and milled rice, from the point of initial production until they are in the hands of the ultimate consumers at the right time, in the right place and at a profit margin. Although majority of the rice producers in Nigeria are smallholder farmers, most of them are into rice production because rice is a commercial crop, given the increasingly high demand for rice in the country. Marketing of locally produced rice takes place at four levels.

First is selling of rice paddies by farmers at farm gate immediately after harvest. At the farm gate, rice paddies are purchased from farmers by itinerant traders, processing companies and cooperative at a give-away price that hardly covers the cost of production. The second level of rice marketing involves a wholesale trading of milled rice at village markets or rice milling centres. This takes place after primary processing of rice paddies (parboiling and milling) by farm households. The third level consists of moving milled rice from rural to urban markets, while the fourth level encompasses mainly retailing in urban areas (FGD, 2012). Figure 2-5 summaries marketing activities of locally produced in Nigeria.

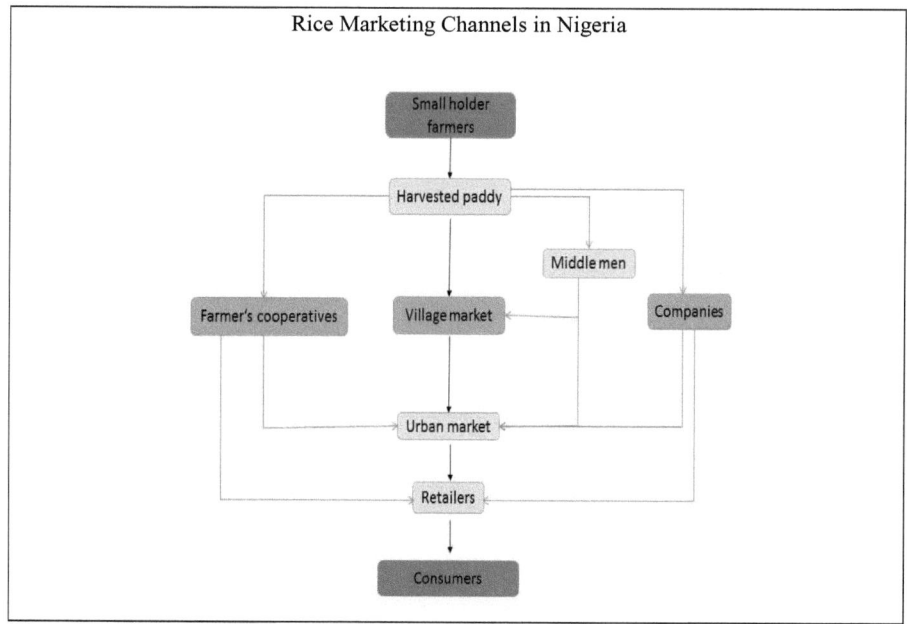

Figure 2-5: Rice marketing channels in Nigeria
Source: Field survey, 2011

Importation of rice into the country creates a sort of competition between locally produced rice and imported rice. However, the locally produced rice is reputed for its peculiar taste and smell compared to imported polished rice. Market structure refers to those characteristics of the market organization that are likely to affect the behaviour and performance of firms such as the number of sellers, the extent of knowledge about each other's action, the degree of freedom of entry and the degree of product differentiation (Lipsey and Steiner, 1981). Although local rice marketing cannot be said to be perfect or pure competition in the strict sense of the word, rice market structure in Nigeria can be described as atomistic competition.

Chapter Three

Literature Review

Introduction

This chapter presents a review of previous studies on adoption and diffusion of new agricultural technologies, impact of technology adoption as well as smallholder market participation. The identified research gaps addressed by the present study are highlighted at the end of each sub-section.

3.1 Paradigms of agriculture development

The role of agriculture as precursor to the acceleration of industrial growth is well documented in the literature. For instance, the historical industrial revolution in England dating back to 1750 and that of Japan in 1880 were both linked to agricultural revolutions (Bezemer and Headey, 2008). In recent time, the success stories of how the Green Revolution kicked-off industrialization in certain developing countries through rapid productivity growth are also noteworthy (Liption 1988).

However, these days, the role of agriculture in development especially in developing countries of the world transcends the basic support for industrial revolution. Therefore, the classical paradigm which prevailed in economic thoughts in the earlier times needs to be broadened, as agricultural growth is capable of accelerating GDP growth at the early stages of development, reducing poverty and vulnerability, narrowing rural-urban income disparities, releasing scarce resources such as water and land for use by other sectors and delivering a multiplicity of environmental services (de Janvry, 2010).

The underperformance of the agricultural sector in Sub-Saharan Africa can be attributed to a myriad of problems originating from lack of well-articulated development policy and underinvestment in the sector by most governments in the region. However, African leaders have renewed their commitments to developing the agricultural sectors in their respective domains with the hope of replicating the Green Revolution experience of Asia in Africa.

As indicated earlier, the Head of States of the African countries agreed to set aside at least 10% of their respective countries' annual national budget for agricultural and rural development (Maputo 2003 declaration on agriculture and food security). The agenda is being coordinated by the Comprehensive Africa Agriculture Development Programme (CAADP) to ensure effective monitoring. CAADP's major pillars include investment in agricultural research to develop improved agricultural technologies and infrastructural development. As a follow up to Maputo 2003 declaration, another declaration was made in Malabo in 2014 to address Africa's growing dependence on foreign markets for food security arguably due to changes in the consumption patterns of most African nations. It has therefore become imperative for African nations to develop its agricultural sector in order to achieve a sustainable development which cut across all facets of the economy and become competitive in the global markets.

Schultz (1964) (as quoted by Ruttan and Hayami, 1972) proposed "High Input Pay-off model" for transforming traditional agriculture into a productive sector. Ruttan and Hayami (1972) noted that agricultural policies based on Schultz's model are capable of generating sufficiently high rate of agricultural growth and stimulating overall economic development. The model, premised on development of improved technologies, availability of production inputs as well as technology dissemination was the basis of Green Revolution which transformed the agricultural sectors and economies of many Asian and Latin

American countries. Similarly, Green Revolution can be achieved in Africa by quality investment in research to develop improved technologies and dissemination of better farming practices using appropriate communication methods.

3.2 The uptake of agricultural innovation

An innovation is defined as a technological factor that changes the production function of the innovators such that they are able to operate on better frontier compared to non-innovators. Although at the initial stage of innovation uptake, there exist some uncertainties about a new technology, whether perceived or objective. The uncertainties diminish over time through acquisition of experience and information, while the production function itself may change as the innovators become more efficient in the application of the innovation (Feder and Umali, 1993).

There is a general consensus among economists that technology adoption can play a major role in the agricultural development of developing countries because the economic base of most of these countries is dominated by agriculture and livelihood of majority of their inhabitants depends on farm output. However, despite considerable research and attention directed to the issues of technological adoption, there seems not to be a consensus on social and economic conditions leading to why some farmers adopt new technology while others do not. However, sociologists and economists have provided various theoretical and empirical explanations for technology adoption.

According to Rogers (1995, 2003), there are five perceived attributes of an innovation. First is the relative advantage of the new technology. This is the degree to which the innovation supersedes current practices in terms of economic returns, social prestige and satisfaction. To put in simple terms, it measures whether an individual perceives the innovation as advantageous. The greater the perceived relative advantage of an innovation, the more rapid its adoption will be. Second is the compatibility of the innovation with the existing values, past experiences and needs of the potential adopters.

An innovation that is compatible with the values and norms of a social system will be adopted rapidly than an innovation that is incompatible. The adoption of an incompatible innovation often requires prior adoption of a new system which may be a relatively slow process. An example of an incompatible innovation is the use of contraceptive methods in countries where religious beliefs discourage use of family planning, as in certain Muslim and Catholic nations.

Third, the complexity or the degree to which an innovation is perceived as difficult to understand and use. The more complex an idea is perceived to be, the longer it will take for it to be adopted. Fourth is the degree to which an innovation has to be experimented on a limited basis. New ideas that can be tried on an installmental plan will generally be adopted quickly than innovations that are not divisible. Finally, observability is the degree to which the results of an idea are visible. The easier it is for people to actually see the results of an innovation, the faster they will accept the idea.

3.2.1 Adoption of agricultural technology

Adoption can be measured as the proportion of land area cultivated with a new technology over total cultivated area. This definition can be referred to as "continuous measure" or intensity of adoption. This measure is mostly applied to situations where a new technology is adopted partially. On the other hand, adoption can be measured in discrete state with binary indicator (of either farmer adopts a technology or not). For example, a farmer may be defined as an adopter if he or she is found to be growing a high yielding variety. Thus, a farmer may be classified as an adopter and still grow some local crop varieties (Doss, 2006).

As pointed out by Feder and Umali (1993), adoption of a new technology can be studied at two levels; at micro level, individual decision unit (farmer) decides whether or not to adopt a new technology, and the intensity of use if adopted. At macro level, aggregate adoption (diffusion) occurs among members of a population over a period of time

Several empirical studies on agricultural technology adoption focused on the uptake of specific technologies (e.g., fertilizer, pesticide and improved varieties), (Shakya et al. 1985; Ransom, et al. 2003; Feleke and Zegeye 2006; Ojiako et al. 2007). For example, Shakya et al. (1985) investigated factors influencing the adoption of modern rice varieties and fertilizer in Southeastern Nepal. Probit regression model was employed to examine discrete choice of technology adoption decision, while Tobit model was used to examine intensity of adoption. Their findings indicated that irrigation, tenure status and access to credit were significantly related to varietal adoption, while household and farm size as well as operator's education was not. Similar variables in addition to fertilizer price influenced the probability of fertilizer adoption and use rates.

In the same vein, Ransom, et al., (2003) employed Tobit regression model to examine determinants and intensity of adoption of improved maize varieties in Nepal. They noted that ethnic group, fertilizer use, off-farm income and extension education positively affected adoption of improved varieties. On the contrary, lack of improved seeds and inadequate knowledge of the new production technologies were major constraints to adoption of the improved varieties.

Given the spate of environmental degradation, climate change and water resources, a number of studies investigated the adoption of soil conservation, sustainable land use and water resources technologies (Norris and Batie, 1987;

Baidu-Forson, 1999; Sidibé, 2005; Amsalu and De-Graaff, 2006). In this direction, Baidu-Forson (1999) investigated the adoption of land-enhancing technologies such as half-crescent shaped earthen mounds and improved 'tassa' in the Sahel zone of Niger. Tobit regression model was employed in order to simultaneously estimate determinants of technology adoption as well as extent or intensity of adoption. They found that a higher percentage of degraded farmland, extension education, lower risk and profitability positively influenced adoption of land-enhancing technologies and the intensity level.

Using Probit regression model, Sidibé (2005) analyzed determinants of adoption of soil and water conservation measures (i.e., zaï and stone strips techniques) in Burkina Faso. Their results indicated that the most significant variables for adoption of both of these conservation practices were training and small ruminants holding. While variables such as education and perception of soil degradation were determinants only for adoption of zaï technique. Membership of farmers' association and area cultivated were positively related to adoption in the case of stone strips.

At times, farmers choose to discontinue the adoption of certain technologies for economic and non-economic reasons. Amsalu and De-Graaff (2006) examined determinants of adoption and dis-adoption of stone terraces as soil and water conservation methods in Ethiopia. Assuming that the decision to adopt the technologies may be different from that of its discontinuity, a bi-variate Probit model was employed. Their investigations revealed that factors influencing adoption and continued use of stone terraces were different. Adoption was influenced by farmer's age, farm size, perceptions on technology profitability, slope, livestock size and soil fertility, while the decision to dis-continue the use of the practice was influenced by actual technology profitability, slope, soil fertility, family size, farm size and participation in off-farm work. However, farmer's perceptions of erosion problem, land tenure security and extension contacts showed no significant influence.

While a significant number of empirical studies mentioned above focused on specific and individual technologies, agricultural technologies are often introduced as a package consisting of several distinct but interrelated components, which if adopted together give better results than when only one component of the technology is adopted. Ersado et al. (2004) examined the adoption decision of high-yielding varieties as well as bands and terraces soil conservation measures in Ethiopia using a Multinomial Logit model. Their findings confirmed that adoption was characterized by sequential adoption pattern and as such, stepwise dissemination of technologies should be encouraged. Byerlee and Pulcano (1986) demonstrated that farmers in Mexico adopted improved varieties, fertilizer, and herbicides in a step-wise manner rather than as a package by fitting logistic diffusion curves of cumulative adoption levels.

Using Bayesian approach, Leathers and Smale (1991) provided plausible explanations for sequential adoption decisions of a technology package by smallholder farmers in the developing countries. Given that a typical farmer is faced with uncertainties about profitability and yield of a new technology due to lack of complete information, he may choose to adopt part of the package while he learns more about the whole package through own experimentation and learning from peers. He then updates his belief on the technology package following Bayes rules. They also demonstrated that the farmer might choose to eventually adopt the whole package when risks and uncertainty were reduced through learning.

Other studies have investigated the role of risk, uncertainty and learning in technology adoption. Abdulai et al. (2008) examined adoption of crossbred cow in Tanzania in the presence of uncertainty. They found that information acquisition and adoption decisions were made jointly. Their findings showed that human capital and scale of operation positively and significantly affected the decision to acquire information and to adopt the technology, while liquidity constraints negatively impacted on the decision to adopt as well as extent of adoption. Furthermore, risk was found to exert a significant effect on adoption through perceived profitability of the new technology. Mara et al (2003) examined roles of risk, uncertainty and learning in adoption of agricultural technologies. They noted that farmer's perceptions and attitudes about riskiness of a technology, as well as trialling and learning played significant roles in adoption decisions.

Cameroon (1999) investigated the role of learning in adoption of high-yielding cotton variety in India. His findings established that learning is an important variable in the adoption process and concluded that learning-by-doing or by-using played an important role in the adoption decision. In the same manner, Conley and Udry (2010) investigated the role of social learning in diffusion of new agricultural technology in Ghana. They found that farmers adjusted their inputs to align with those of their information neighbours who were surprisingly successful in the previous periods. They also found that farmers increased (decreased) input use when an information neighbour achieved higher than expected profits when using more (less) inputs than they previously used. Further evidence of learning was provided by changes in profits that corresponded to input changes that appeared to be mistakes and those that appeared to be correct, subject to a conjecture regarding the optimal level of input use. Thus, learning implied that farmers respond to both signal and noise particularly in the early stages of adoption process.

Generally, policy, infrastructure and institutional factors (such as extension service, credit and access to input and markets), farm level factors (such as soil characteristics, risk and uncertainty), individual producers characteristics (e.g assets, education, perception and learning), as well as participation in social networks (e.g learning from peers and association) appear to play significant roles in adoption of a new technology. However, most of the empirical studies described above are limited to investigation of determinants of adoption using binary and limited dependent variable models. These models do not explicitly address the effects of regressors on the time-path of adoption, which is an important attribution of the adoption process. Thus, diffusion and longer term adoption dynamics remain unexplored.

The few studies conducted so far on adoption of NERICA technology include Diagne (2006), who investigated the determinants of adoption and estimates of actual and potential adoption rates of NERICA varieties in Côte d'Ivoire using standard binary and IV Poisson regression. He reported that growing rice partially for sale, household size, age, farming experience in growing of upland rice and participation in PVS trials had positive and statistically significant impact on NERICA adoption, while having a secondary occupation exacted negative impact. He argued that predicted probability of adoption would be biased if non-exposure to technology was not taken into consideration, given that the technology was newly introduced and only few farmers had knowledge of its existence. As such, the predicted probability of (actual) adoption was found to be 4%, whereas the potential adoption rate was 27% if non-exposure was taken into consideration. Tiamiyu et al, (2009) employed Tobit model to examine determinants and intensity of NERICA adoption in Nassarawa and Kaduna States of Nigeria. They found that education, extension visits, farming experience, land ownership, credit usage and level of commercialization positively influenced uptake and intensity of NERICA adoption.

Following the work of Diagne (2006), Ojehomon et al. (2012), Nguezet et al. (2013) and Dibba et al. (2012) employed IV Probit to examine probability of adoption of NERICA varieties in Nigeria and the Gambia respectively. In the case of Nigeria (Ekiti, Osun, Niger, Kano States), access to extension, having farming as primary occupation, living in a PVS village and number of years of residence in rice producing community had positive impact on adoption, while education and family size had negative impact on the probability of adoption. Whereas in the case of the Gambia, living in a village where PVS had been implemented, number of years of schooling and contact with extension agents had positive and significant effect on adoption while cultivation of lowland rice had negative effects. Nevertheless static models such as Tobit, Probit and Poisson models employed in their studies can only provide information on innovation uptake at a point in time but lack credibility for longer term adoption dynamics due to changing economic factors and time-varying covariates.

3.2.2 Diffusion of agricultural technology

As indicated above, diffusion of agricultural technology is the cumulative processes of adoption measured over successive time periods. Given that new technologies offer opportunities for increasing productivity, improving product quality and incomes, what determines the actual improvements in productivity and product quality is not the rate of development of the new technologies, but the speed and extent of their application in commercial operations (Stoneman and David, 1986). In essence, the extent and speed of the spread of an innovation within the producers' population are important to technological change and increasing productivity at the aggregate level.

Experience has shown that several factors such as lack of credit, limited access to information and inputs, as well as inadequate infrastructure can constrain technology diffusion. However, the nature and intensity of the impact of these constraints may vary according to the type of technology. To overcome some of these constraints, policy makers have generally pursued two general strategies: information provision (for example, extension programs) and the provision of subsidies and support programs (inputs, and credit subsidies, the provision of complementary infrastructure, and risk-reducing programs). However, the effectiveness of these strategies in fostering technology adoption have been issues that have drawn considerable attention (Feeder and Umali, 1993).

As posited by Rogers (1995, 2003), diffusion is the process through which (1) an innovation (2) is communicated through certain channels (3) over time (4) among the members of a social system. While the characteristics of an innovation as perceived by members of a social system are important in the diffusion process, the method of communicating the innovation to potential users has been an interesting area of research. Although mass media channel appears to be effective in creating general awareness about an innovation, interpersonal channels can be more effective in forming and changing attitudes towards an innovation, and therefore important in influencing the decision to adopt or reject a new idea. Most individuals evaluate an innovation not on the basis of scientific research by experts, but through the subjective evaluations of near-peers who have adopted the innovation.

Time factor may also underscore the rate of technology adoption in terms of relative speed with which an innovation is adopted by members of a social system. Rate of adoption is usually measured as the number of members of a system that adopt the innovation in a given time period. As shown previously, an innovation's rate of adoption is influenced by the five perceived attributes of an innovation. If calibrated on time horizon, innovativeness is the degree to which an individual or other unit of adoption is relatively earlier in adopting new ideas than other members of a social system. A social system is defined as a set of interrelated units that are engaged in joint problem-solving to accomplish a common goal. The members or units of a social system may be individuals, informal groups, organizations, and/or subsystems. However in a given social system, a change agent is an individual who attempts to influence clients' adoption decisions in a direction that is deemed desirable by the change agency.

Rogers (1995, 2003) classified members of a social system into five adoption categories, which are innovators; early adopters, early majority, late majority and laggards. Innovators are willing to take risks, have the highest social status, have financial liquidity, and they have closest contact with scientific sources and other innovators. These attributes give them an edge over other members of the social system and as result, they are able to adopt a new technology faster. Early adopters are individuals with higher degree of opinion leadership, higher social status, and financial liquidity, advanced education and are more socially forward than late adopters. They are more discreet in adoption choices than innovators.

The early majority category adopt an innovation after a varying degree of time that is significantly longer than the innovators and early adopters. This category has above average social status, contact with early adopters and seldom hold positions of opinion leadership in the social system. While the late majority approach an innovation with a high degree of scepticism after the majority in society has adopted the innovation. Finally, the last category of adopters in a social system is referred to as laggards. Unlike some of the previous categories, individuals in this category show little or no opinion leadership. These individuals typically have an aversion to change. Laggards typically tend to be focused on "traditions", they have lowest social status, lowest financial liquidity, oldest among adopters and in contact with only family and close friends (figure 3-1).

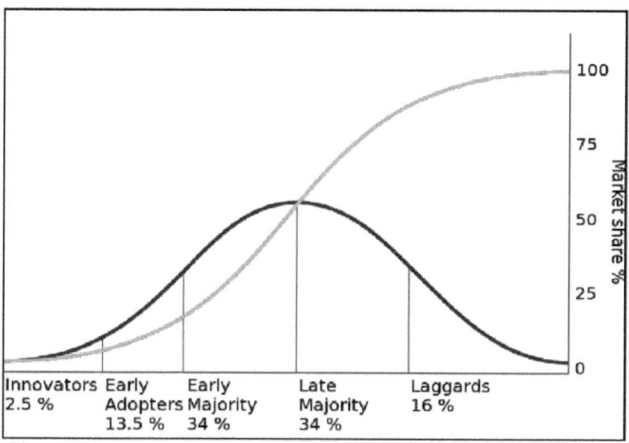

Figure 3-1: Categorization of innovation adopters.
Source: Rogers (1995).

As indicated above, empirical studies on individual adoption behavior have received considerable attention in the literature, while diffusion of agricultural technologies amongst smallholder farmers particularly in Africa has not been given adequate attention. Many theoretical and empirical frameworks have been developed to explain the process of innovation diffusion within a social system. The epidemic diffusion theory, which laid foundation for empirical modeling of innovation diffusion described it as a disequilibrium process resulting from information asymmetries among potential adopters. As demonstrated differently by Griliches (1957) and Mansfield (1961), information about the existence of a new technology is spread by direct contact between a potential user and a user who has adopted and successfully used the technology. This generates a time path of diffusion that assumes an "S shaped" form where the speed of diffusion is based on frequency of contact (figure 3-2).

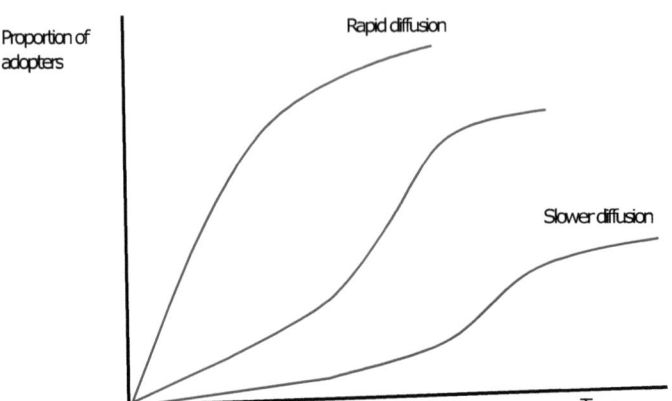

Figure 3-2: Logistic or sigmoid diffusion curve
Source: Adopted from Griliches (1957) and Baptista, (1999)

Assuming a population of *n* potential adopters, and defining $m(t)$ as the number of individuals who have adopted the technology at *t*, the basic epidemic model can be expressed as:

$$dm(t) = \beta [m(t)] \cdot [n - m(t)] \, dt \qquad (3\text{-}1)$$

where $\beta > 0$ is the parameter defining speed of adoption. The number of non-users adopting the technology in a period increases as the proportion of users in a social system increases. This is because as information and experience about the new technology are accumulated over time, it becomes less risky for non-users to adopt the technology. Besides, stock and epidemic effects resulting from increase in the number of adopters can cause bandwagon effects. However, epidemic diffusion theory is criticized based on assumptions of fixed and homogenous population of potential adopters, who are described as passive recipients, rather than active information seekers.

However, alternative views have approached the subject of diffusion from an equilibrium framework. These models concentrate on the decision-making process of individual firms, as the uncertainty models do, but assume that information in the economy is perfect. Hence, knowledge about the characteristics and profitability of the new technology is widespread. Differences in adoption timing happen because potential adopters differ from each other. This heterogeneity is represented by one or more key characteristics that are assumed to be crucially important in determining responsiveness to the technology (Baptista, 1999).

The Games Theory application to innovation diffusion postulates that firms exhibit strategic behaviour in a real world, even if they are identical and there are no risks and uncertainties about a new technology. The strategic behaviour involves being able to decide on the optimum time to adopt an innovation so as to be ahead in competition (Sarkar 1998). The basic premise of the model is that adopters can be classified as innovators or as imitators and the speed and timing of adoption depends on their degree of innovativeness, and imitation or learning from successful adopters (Bass, 1969).

However, Karshenas and Stoneman (1993) synthesized recent theoretical advances in diffusion studies into rank (or probit), stock (or game theoretic), and order effects. According to them, the essential prediction of the diffusion theories is that potential adopters of a new technology have different (preferred) adoption dates due to firm's heterogeneity and preferences arising from rank, stock, and order effects. Rank effect suggests that potential adopters have different inherent characteristics (such as firm size) and as a result, obtain different returns from the use of a new technology – these expectations generate different preferred adoption dates.

The stock effect shows that the benefit to a marginal adopter decreases as the cumulative number of adopters increases; this is because as the number of accumulated adopters increases, a point is reached when they have a large impact on the market so that adoption is no longer profitable. While order effect shows that returns to a firm from adopting a new technology depends upon its position in the order of adoption. For instance, farmers who have access to exceptional information, skills or resources are classified as higher order adopters. Higher order adopters obtain greater returns than low order adopters; this may motivate their decisions to adopt earlier.

Gourlay and Pentecost (2002) applied Karshenas and Stoneman (1993) model to diffusion of ATMs (Automated Teller Machines) in the United Kingdom, they found that rank and order effects significantly influenced diffusion of ATMs in the UK. Similarly, Abdulai and Huffman (2005) applied the model to diffusion of crossbred cow in Tanzania. Their findings indicated that rank, stock and order effects played significant roles in the diffusion process of crossbred-cow technology. Specifically, rank effects such as education of the household head, herd size and distance to the nearest local markets were found to have significant effects on hazard of technology adoption.

Given that NERICA technology offers opportunities for increased rice productivity and food security in Nigeria, the present study analyzes impact of farm and non-farm factors on the duration waited by farmers before adopting the technology using a combination of Karshenas and Stoneman (1993) optimal time framework and duration analysis. Duration analysis bridges the gap between adoption and diffusion studies, by applying cross-sectional and time-series data jointly in a dynamic framework. It is therefore relevant in explaining not only technology diffusion but also what factors influenced the observed time patterns of adoption (Dadi et al. 2004).

Duration or survival analysis originated from biomedical sciences and industrial engineering where it is applied to analysis of time duration until events happen, such as death of biological organisms and failure in mechanical systems, provided that the object is at risk. Duration analysis has been widely employed in social sciences, especially in modeling duration of unemployment in labour economics and survival of institutions. In recent time, there is increasing application of the method to examine adoption of agricultural technologies, although most of the existing studies focused on diffusion of farming systems and process innovations.

For example, Burton et al. (2003) applied the model to adoption of organic horticulture in the UK. Similarly, D'Emden et al. (2006) studied the adoption of conservation agriculture amongst Australian farmers, while Murage et al. (2011) examined adoption of push-pull technology for Striga control in Kenya. These studies showed factors that prompted farmers to adopt agricultural innovations as well as the speed of adoption. Particularly, they highlighted the roles played by time dependent variables such as change in price of the technology in the uptake and timing of adoption of agricultural innovations, which cannot be accommodated by static models. The present study employs duration analysis to examine adoption and diffusion of product innovation amongst smallholder farmers which has been largely omitted in the literature.

A discrete-time duration model is employed in the study because although agricultural innovation adoption occurs in continuous time, the economic data on technology adoption are usually available only on yearly basis, with the precise time of adoption within the yearly interval not known with certainty (i.e., the exact month and day are not known as in engineering and biomedical events). Such data are referred to as grouped, banded or interval-censored data. A discrete-time duration model has been found to be most appropriate for estimating hazard probabilities in such cases (Burton et al., 2003). Moreover, discrete-time duration models offer theoretically consistent approaches for incorporate time-varying covariates and flexible specifications of duration dependence in duration models than the continuous time models (Jenkins 1995).

3.3. Impact evaluation

Evaluating the impact of an innovation, for example the NERICA technology, is important in examining its effectiveness such as on-farm performance and yield characteristics and consequently, its contributions to welfare of smallholder producers. Although, researchers have employed various econometric approaches to examine the causal effect of agricultural technology on various outcomes of interest, yet there are questions on credible approaches to construct suitable counterfactual situation in order to identify the true causality of technology adoption. While initial empirical studies on causal effects of programs and policies have focused on the use of traditional econometric methods for dealing with endogeneity (such as fixed effect methods and instrumental variables approaches), subsequent works have employed insights from the semi-parametric literature to develop new estimators for a variety of settings, requiring fewer functional form and homogeneity assumptions.

Rubin Causal Model (RCM), popularly known as potential outcomes model has been of particular importance in this direction (Imbens, 2009). As stated in Donald Rubin (1974, 1977, and 2004), causal effects are comparisons of the potential outcomes that would have been observed under different exposures of *units* to *treatments*. For example, if $Y(0)$ is the outcome of an event without treatment, and $Y(1)$ is the outcome with treatment, given that Stable Unit Treatment Value (SUTVA) assumption holds, i.e., (a) units do not interfere with each other and treatment applied to one unit does not affect the outcome for another unit, and (2) there is only a single version of each treatment level and potential outcomes are well defined. The difference, $Y(1) - Y(0)$, is an obvious definition of the causal effect of the treatment.

Based on Rubin's theorems, the true impact of adoption of NERICA varieties can be obtained by comparing the observed outcome of adoption to the outcome that would have resulted, if the adopters had not adopted the technology (or if non-adopters had adopted the technology), i.e a comparison of the factual and counterfactual outcomes. Otherwise, it will be difficult to conclude that the outcome realized by the adopters is due to adoption of the technology.

Three parameters are most frequently estimated in the literature as measures of mean impact of treatments on the desired outcome variables. The first one is the population average treatment effect (ATE), which is simply the difference between the expected outcomes after participation and nonparticipation: $ATE=E(\tau)=E[Y(1)-Y(0)]$. This parameter answers questions about the mean or expected effect on the outcome if individuals in the population were randomly assigned to treatment. A more prominent evaluation parameter is the so-called average treatment effect on the treated (ATT), which is given by $ATT=E(\tau|D=1)=E[Y(1)|D=1]-E[Y(0)|D=1]$. The expected value of ATT is defined as the difference between expected outcome values with and without treatment for those who actually participated in treatment. The third parameter of interest is local average treatment effect (LATE). LATE is the average treatment effect for individuals whose treatment status is influenced by changing an exogenous regressor that satisfies an exclusion restriction. It is the mean treatment effect among the compliers, i.e., unit that received treatment if and only if induced to do so by an instrumental variable (Imbens and Angrist 1994).

The econometric procedures for estimation of casual effect of a treatment on the outcome of interest can simply be classified into two groups; experimental and non-experimental approaches. The non-experimental approaches are often referred to as quasi-experimental methods because they are designed to mimic real experimental situations. In randomized experiments, assignment into treatment and control groups is random, so the control group has the same distributions of both observed and unobserved characteristics as the treatment group and as such, the control group provides a suitable counterfactual which makes it straightforward to obtain estimators for the average effect of treatment on the outcome variable.

Observable characteristics refer to factors such as age, education and location variables which are measurable to a researcher while unobservable factors include the innate managerial and technical abilities of the subjects which cannot be simply detected (e.g., motivation, innate managerial abilities and intelligence). Since assignment to treatment is random in an experimental set up, individuals assigned to treatment and control groups differ in expectation only through their exposure to treatment thus, randomization solves the problem of selection bias (Smith and Todd, 2005). Albeit, there could be sample selection problems in randomized control trials if factors other than random assignment influence program allocation. For example, if a parent moves his child from a class (or a school) outside of a program (i.e., control group) to a school within the program (i.e., treatment group) (Duflo and Kremer 2005).

Other concerns on randomised experiment include external validity, ethical issues, partial or lack of compliance, selective attrition, and spillover effects. External validity is concerned with how to replicate or generalize the results obtained through randomized evaluations especially if the geographical location and environmental settings of the experiment is different from where it is replicated. At times, the people who agree to participate in an experiment (as either experimental or controls units) are not themselves randomly drawn from the general population so that, even if the experiment itself is perfectly executed, the results are not transferable from the experimental to the parent population and will not be a reliable guide to policy in the parent population.

Similarly, withholding a particular treatment from a random group of people and providing access to another random group of people may be simply unethical. Compliance may also be a concern in randomised experiments. For instance, if a fraction of the individuals who are offered treatment fail to accept assignment or some members of the control group receive the treatment. This situation is referred to as partial (or imperfect) compliance (Duflo and Kremer 2005; Khandker et al. 2010; Deaton, 2010). Finally, potential spillover effects may arise when treatments are indirectly received by the control group, thereby confounding the estimates of the program's impact. For example, people outside an experimental sample may move into a village where RCT is being implemented thus, contaminating the programs effects (Khandeker et al. 2010).

However, when the data available for a study are from a cross-sectional survey or observational studies, as in the case of the present study, there would not be information on the counterfactual situation because it is practically impossible to observe outcomes for an individual in the two states (i.e., factual and counter factual situations). At the same time, we cannot simply use a non-treatment group as control group due to self-selection problems. In such cases, quasi-experimental approaches have been found very useful.

One of the notable quasi-experimental approaches widely used in the literature to analyse observational data is Heckman's (1979) sample selection model. Heckman's sample selection model triggered both a rich theoretical discussion on modelling selection bias. Heckman's model is estimated in two stages; in the first stage, the endogenous variable representing participation is modelled. This is done by estimating the probability (propensity scores) of a participant being in the treatment group or otherwise as indicated by dummy variable representation. Then, the estimated propensity scores are used to calculate correction factor (otherwise known as Millis ratio). In the second stage, the unobserved selection factors are treated as problem of specification error or a problem of omitted variable which is corrected by incorporating consistent estimation of the regression model (Shenyang-Guo and Fraser 2010).

Using a 2006 household survey in Mali, Gubert et al. (2010) employed Heckman two-step procedure to evaluate impact of financial remittances by relations living in cities on incidence of poverty amongst farm households in rural Mali. They found that remittances reduced poverty rates by 11% and the Gini coefficient by about 5%. They also found that households in the bottom quintiles were more dependent on remittances, which are less substitutable by additional workforce. Similarly, Bacha et al. (2009), investigated the poverty reduction impacts of adoption of small-scale irrigation technology in the Ambo district of western Ethiopia using Heckman's selectivity model. Their results indicated that the incidence, depth, and severity of poverty were significantly lower among farm households with access to irrigation. In addition to irrigation, other variables such as farm size, livestock holding size, land productivity and family size significantly influenced the level of household consumption expenditure.

However, a major drawback of Heckman's model is that it can only corrects selectivity bias arising from unobservable characteristics. Besides, the model imposes normality assumption and proper identification to generate credible estimates. In contrast to Heckman two-stage procedure, the Instrumental Variable (IV) approach is usually employed when a variable can be identified that is related to participation but not outcomes. This variable is known as 'instrument' and it introduces an element of randomness into assignment which approximates the effect of an experiment (Imbens and Angrist, 1994). Where it exists, estimation of the treatment effect can be done by using a standard instrumental variables approach. Where variation in impact of treatment across people is not correlated with the instrument, the IV approach recovers an estimate of impact of treatment on the treated (ATT). However, if the variation is related to the instrument, the parameter estimated is *Local Average Treatment effect* (LATE). The main drawback to the IV approach is that it is often difficult to find a suitable instrument because to identify the treatment effect, one needs at least one regressor which determines programme participation but not the outcomes in the model (Imbens and Angrist, 1994; Bryson et al. 2002).

The empirical studies on impact of adoption of NERICA technology have all employed IV approach. Adekanbi et al. (2009) utilized a 2004 household survey data to examine the impact of adoption of the New Rice for Africa (NERICA) varieties on income and poverty status of rice producers in Benin Republic using non-parametric Wald estimator and Instrumental Variable approach to estimate *Local Average Treatment Effect* (LATE). They opined that since NERICA varieties were introduced through participatory varietals selection (PVS) approach, only few farmers who participated in PVS were possibly exposed to the technology. Exposure to NERICA was identified as an instrument that correlated with adoption but not the outcome. Their results indicated that the adoption of NERICA varieties had a positive and significant impact on household expenditure and poverty reduction amongst rice farming households in the study area.

Similarly, Nguezet et al. (2011) evaluated the impact of the New Rice for Africa (NERICA) varieties on income and poverty status of the rice producers in Nigeria using the Wald estimator and Instrumental Variable (IV) to estimate *Local Average Treatment Effect* (LATE). They also found that technology adoption had positive impact on income and reduce poverty incidence amongst the technology adopters. However, non-exposure to technology may not be a valid instrument considering massive dissemination campaign of NERICA technologies and unrestricted access to NERICA seeds. Even if exposure to the technology was a valid instrument, Deaton (2010), and Heckman and Urzua, (2009) strongly argued that LATE is often not the causal estimand of interest because it only provides information on treatment effect for the subsample identified by instruments. In essence, the problem of identification and exogenity of instruments call LATE estimates to question and often times, these estimates may be difficult to interpret in providing answer to an interesting economic question.

Another quasi-experimental approach employed in impact evaluation is the non-parametric method of matching. Traditional matching estimators pair each program participant with an observably similar non-participant and interpret the difference in their outcomes as the effect of the program. Matching assumes that sample selection can be explained strictly by observables characteristics. When selection bias is only due to observables, matching is a useful tool to estimate treatment effect. The most attractive feature of matching compared with the regression type estimators is that matching neither imposes functional form restrictions such as linearity nor assumes a homogeneous treatment effect in the population. Covariate matching as the basis of correcting for bias due to observables is intuitive, since the source of the bias is the difference in observables in the treated group and comparison group. Matching on covariates by definition will remove this difference and hence the bias. However, when there are many covariates, it is impractical to match directly on covariates because of the curse of dimensionality (Zhao, 2003).

However, Rosenbaum and Rubin (1983) proposed the use of the propensity scores as the basis for matching treatment with control group, reflecting the probability of being in either of the groups, conditional on their different observed characteristics. Nonetheless, the validity of the matching methods depends on two main assumptions. The first is if one can control for observable differences in the characteristics between the treated and non-treated groups, the outcome that would result in the absence of treatment is the same in both cases. This identifying assumption for matching is known as the Conditional Independence Assumption (CIA). CIA allows the counterfactual outcome for the treatment group to be inferred and for any difference between the treated and non-treated groups to be attributed to the effect of the programme. To fulfil the CIA assumption, a very rich dataset is required since the evaluator needs to be confident that all the variables affecting both participation and outcome are observed.

That is, it is assumed that any selection on unobservables is trivial in that unobservables do not affect outcomes in the absence of treatment. Where data do not contain all the variables influencing both participation and the outcome, CIA is violated since the programme effect will be accounted for in part by information which is not available to the evaluator. Second is common support assumption which states that there must be an overlap in propensity scores across the treatment and control groups (Bryson et al. 2002; Shahidur et al. 2010).

Given that these assumptions are fulfilled and matching has been carried out, the average treatment effect of the program is then calculated as the mean difference in outcomes across these two groups. As documented by Caliendo and Kopeinig (2008), matching estimators contrast the outcome of a treated individual with outcome of the comparison group members. PSM estimators differ not only in the way the neighbourhood for each treated individual is defined and the common support problem is handled, but also with respect to the weights assigned to these neighbours.

The commonly used matching estimators are; (I) Nearest Neighbour (NN) matching which is done by matching individual from the treated group having similar propensity scores with another individual with similar characteristics in the control group. (II) Caliper and Radius matching is implemented such that an individual from the comparison group is chosen as a matching partner for a treated individual that lies within a certain caliper ('propensity range') and is closest in terms of propensity score. (III) Stratification and Interval (SI) matching partitions the common support of the propensity score into a set of intervals (strata) to calculate the impact within each interval by taking the mean difference in outcomes between treated and control observations. This method is also known as interval matching, blocking and sub-classification. (IV) Kernel matching (KM) and local linear matching (LLM) are nonparametric matching estimators that use weighted averages of the propensity scores of individuals in the control group to construct the counterfactual outcome.

Mendola (2007) employed the propensity score matching approach to study impact of adoption of high yielding rice varieties on the economic wellbeing of smallholder farm households in rural Bangladesh. He found that adoption of agricultural technology directly enhanced productive capabilities and significantly reduced poverty incidence amongst the adopters when compared with the non-adopters. Becerril and Abdulai (2010) employed propensity score matching approach to evaluate impact of adoption of high yielding maize varieties on welfare of farm households in Mexico. They found that technology adoption enhanced productivity of the adopters, raised their per-capital expenditure and reduced poverty incidence amongst them. Specifically, their empirical findings showed that technology adoption helped raise household per-capital expenditure by an average of 136 -173 Mexican pesos, thereby reducing the probability of the adopters falling below the poverty line by 27-31%.

Similarly, Kassie et al. (2011) evaluated impact of adopting improved groundnut varieties on the wellbeing of the producing households in rural Uganda using propensity score matching. They found that adoption of high yielding groundnut significantly increased crop income and enhanced the chances of escaping poverty by technology adopters. However, a major drawback of the PSM approach is that of fulfilling the Conditional Independence Assumption (CIA). CIA implies that once observable characteristics are controlled for, technology adoption is random and uncorrelated with the outcome variables. However, there may still be some systematic differences between adopters' and non-adopters' outcomes even after conditioning on observable characteristics, due to unobservables (Smith and Todd 2005). Another issue is that PSM requires large samples with substantial overlap between the treatment and control groups for meaningful implementation (Bryson, 2002).

The *Before-and-After* estimator otherwise known as *difference-in-difference* method is another quasi-experimental approach which has been widely used in assessing impact of treatment on the outcomes of interest. The approach is commonly referred to as difference in difference or natural experiment, and it operates by comparing the outcomes of a group of individuals in the treatment group with outcomes of the same group before receiving

treatment. The approach removes selection bias due to observable characteristics and macro effects by common differencing while ATT is calculated as the difference between the outcome before and after the treatment. However, it relies on two critically important assumptions of common time effects across groups and no composition changes within each group. Together, these assumptions make choosing a comparison group extremely difficult (Blundell and Costa-Dias 2000). Feder et al. (2004) evaluated impact of farmer field schools, an intensive participatory training program emphasizing integrated pest management in Indonesia using *difference-in-difference* estimator. The evaluation focused on whether program participation improved yields and reduced pesticide use among the program participants and their neighbors who might have gained knowledge from them through informal communications. The study utilized panel data collected from smallholder farmers in Indonesia from 1991 to 1999, while *difference-in-difference* approach was employed to evaluate impact of participation on yields and pesticide usage. Empirical results showed that the program did not have significant impacts on the performance of graduates from farmer field schools and that of their neighbours.

Omilola (2009) investigated the poverty reduction effects of adoption of irrigation technology in rural Nigeria using ordinary least squares and *difference in difference* approach to correct for endogenity and estimated the unconditional treatment effects of the technology on incomes of the adopters. He found a positive relationship between technology adoptions, income and poverty reduction amongst smallholder farm households in the study area. However, the approach can only be applied to panel data and only corrects for selection on observables. Moreover, the assumptions of common time effects across groups and no compositional changes within each group may sometimes be difficult to fulfil (Blundell and Costa-Dias 2000; Abadie, 2005).

Regression discontinuity (RD) design is another parametric quasi-experimental design which is based on the idea that the sample in the neighbourhood of a cut-off point (above and below) represents features of the randomized experimental design. The approach is applicable if subjects in treatment and control groups are similar in their characteristics but vary only in treatment assignment. As such, a difference in mean outcomes of the treated and control groups gives the impact of an intervention. RD has two versions; the sharp discontinuity designs in which the cut-off point deterministically establishes treatment status. That is, everyone eligible gets the treatment, and no one ineligible gets it. Fuzzy discontinuity design describes a situation whereby assignment to treatment does not require a sharp cut-off. It is applicable in situations where the probability of assignment into treatment and the control groups is different (Shahidur et al., 2010).

Chang (2013) examined the effect of the old farmers' pension program on farm succession in Taiwan using the analytical framework of regression discontinuity design and found that the pension program induced elderly farmers to work more on the farm and less off-farm. Although the primary policy objective of the pension program was to secure the well-being of elderly farmers, an undesired negative effect on farm succession was found. Similarly, Eggleston (2014) also employed regression discontinuity analysis to assess impact of China's new rural pension program on promoting migration of labour and off-farm employment. The results revealed a perceptible difference in household behaviour at the age of pension eligibility as the adult children of the beneficiaries were more likely to take an off-farm job and migrate to other areas in search of such jobs. These abrupt changes in household behaviour at the cut-off suggest that migration can be discouraged if households are credit unconstrained. The striking drawbacks of regression continuity designs are that estimation procedure requires that the functional form of the relationship between the treatment and outcome is correctly modelled. There can also be spillover effects resulting from lack of clearly defined cut-offs (Imbes and Wooldridge, 2009).

However, a quasi-experimental approach that accounts for self-selection on both observable and unobservable factors is Endogenous Switching Regression (ESR) proposed by Lee (1982). ESR is a generalization of Heckman's model (1979) in which sample selection is treated as a problem of specification error or omitted variable, which can be corrected by explicitly using information gained from the selection equation for consistent estimation of the outcome equation (Shenyang and Fraser 2010).

Kabunga et al. (2012) employed the ESR approach to analyze the yield effects of adoption of Tissue Culture (TC) banana among smallholder farmers in Kenya. The results of their econometric estimations revealed presence of negative selection bias, implying that farmers with lower than average yields were more likely to adopt TC technology. Controlling for this bias resulted in a significant net yield gain of 7% for technology adopters. Abdulai and Huffman (2014) also examined factors that influenced the adoption of soil and water conservation technology as well as impact of adoption on yield and net-returns among rice farmers in Northern Ghana by employing the ESR approach to account for selectivity bias. The results showed that there was endogenous switch in technology adoption decision thus, technology adoption may not have the same effect if non-adopters had chosen to adopt the technology. Furthermore, a positive selection bias was observed for both rice yields and net-returns, suggesting that more-productive farmers tend to adopt the technology.

By employing ESR approach, the present study provides information not only on determinants of innovation uptake and the differential impact of the explanatory variables on outcomes of interest for innovators and non-innovators, but it also estimates average treatment effects of innovation uptake. While previous studies that employed ESR have not considered causal effects of technology adoption on poverty incidence amongst smallholder farm households, the present study contributes to knowledge by investigating the determinants of NERICA technology adoption as well as impact of adoption of the NERICA varieties on productivity, net-returns and poverty incidence amongst rice producing households in Nigeria. Furthermore, following Ali and Abdulai (2010), the study also investigates the effect of farm size on net-returns and poverty incidence amongst rice producing households in Nigeria.

3.4 Market participation by smallholder farmers

While promoting agricultural productivity amongst smallholder farmers in the developing countries, facilitating their access to rewarding markets is crucial in translating increased agricultural production to income. Poor market access has been identified as major limiting constraint to agricultural development in the developing world (Renkow et al. 2004; Jayne et al. 2010). In recent time, series of agricultural marketing reforms have been introduced in most of the developing countries of the world to promote efficient marketing systems. At the national level, trade liberalization and many other market reforms policies have been implemented following the mid-1980s Structural Adjustment Programs recommended by the World Bank and International Monetary Fund (IMF).

For instance, until the 1990s most developing countries generally taxed their agricultural sectors whereas, the developed countries generally supported their own agricultural sectors through subsidies to producers, high tariffs, and other nontariff measures such as import restrictions. However, the pattern of incentives has begun to change with the market reforms in the developing countries as export taxes have been eliminated in many cases, average tariffs have declined rapidly and other import restrictions, such as foreign exchange allocations for import, have effectively disappeared (World Bank, 2001; Aksoy and Beghin, 2004).

However, the empirical research on smallholder market participation behaviour has been extremely thin, perhaps especially with respect to staple food commodities and reasons why farm households in the developing countries are not participating in markets are not well understood (Bellemare and Barrett 2006; Barrett, 2008). Renkow et al. (2004) highlighted the role of infrastructural facilities in smallholder market participation. According to them, transportation and communications infrastructure facilitates spatial integration of product and factor markets, reduce transactions costs and promote market participation. A number of studies have examined market participation as a two-stage process of discrete choice of whether to participate in market or not, and continuous choice of the quantity traded. In this direction, Goetz (1992) examined determinants of participation in coarse grain market in Senegal. He found that changes in output prices stimulated production of marketed surpluses. In particular, better information significantly raised probability of market participation of the selling households, while access to coarse grain processing technology significantly influenced quantities sold.

Similarly, Fafchamps and Hill (2005) examined Ugandan smallholder farmers' choice of whether to sell their Robusta coffee beans at the farm-gate, or transport them to market. The study employed standard Probit model to investigate the decision to sell at market. They found that distance to market exert negative impact on probability of market participation while household wealth – measured as the value of all non-land assets of farm household had positive and significant effect on probability of market participation. The second stage regression estimation showed that quantity traded at market increased with proximity to the market. They concluded that wealthy farmers were able to participate in market because they were not liquidity constrained.

Bellemare and Barrett (2006) investigated if market participation decisions (discrete) choice and sales or purchase volumes (continuous) were made sequentially or simultaneously using a sample of smallholder farmers from Ethiopia and Kenya. A two-stage ordered Tobit model was employed in order to examine if market participating decisions were made sequentially or simultaneously. A discrete choice model was estimated in the first stage, while quantity transacted was examined in the second stage. Their empirical results showed that market participation decisions were made sequentially and market participants were indeed more price responsive and less likely to be vulnerable to traders' exploitation.

Ouma et al. (2010) investigated impacts of household, farm, market access, and locational characteristics on the jointly determined banana market participation decisions and transacted quantities of sellers and buyers in Rwanda and Burundi using bivariate Probit model and Heckman's procedure. Their findings showed that market participation decisions were highly influenced by transaction costs. Transportation costs and lack of market information had negative effect on market participation and quantities transacted. Similarly, non-price factors such as land tenure, labour availability, off-farm income, gender, farming experience had strong correlation with the transacted volume. Musara et al, (2011) also analysed determinants of farmers' participation in contract cotton farming in Zimbabwe using Logit regression model. They found that land size, dependence ratio, years of schooling, access to other income and duration of growing cotton influenced farmers' participation in cotton contact farming. They concluded that contract farming could be a reliable source of market and credit to smallholder farmers in Zimbabwe.

However, only few studies have examined impact of smallholder market participation on welfare of smallholder farm households. Fischer and Qaim

(2012) showed that collective marketing by smallholder farmers positively impacted on household income, size of farm holdings and crop productivity amongst banana producers in Kenya. Equally, Lubungu (2013) investigated the welfare effects of participation in cattle markets in Zambia and found that participation in cattle markets led to an increase in household income by as much as 52-64%. The results showed that participation in livestock markets enhanced welfare of smallholder households and contributed to poverty reduction. Nonetheless, both Fischer and Qaim (2012), and Lubungu (2013) employed propensity score matching to examine impact of market participation on the outcomes of interest.

As pointed out earlier, the underlying assumption of PSM approach they employed is unconfoundedness, selection on observables or conditional independence and as such, the approach does not account for selectivity bias that may arise from unobservable characteristics of farm households in the sampled population. The contributions of this study to the literature on market participation are in two folds. First, it provides new insights and empirical results on determinants of market participation by smallholder farmers thereby augmenting the thin literature on market participation. Second, the study examines impact of participation, not only on economic returns on rice farming but also on incidence of poverty amongst farm households, which is hitherto missing in the literature. The study employs an econometric procedure capable of accounting for selection bias that may arise from observable and unobservable characteristics of farm households, given that they self-selected themselves into market participation.

Chapter Four

Conceptual Frameworks and Empirical Models

Introduction:

This chapter presents conceptual frameworks and empirical models employed in this study. Section 4.1 showcases the conceptual framework and empirical models for adoption and diffusion of NERICA technology. Sections 4.2 describes the conceptual frameworks and empirical strategy for determinants of NERICA adoption and its impact on net returns and welfare of the producing households. While section 4.3 presents that of determinants of smallholder market participation and its impact on Return on Investment as well as welfare of rice producing households.

4.1 Conceptual framework for adoption and diffusion of NERICA technology

Following Karshenas and Stoneman (1993) and Abdulai and Huffman (2005), adoption and diffusion of NERICA technology are modeled in an optimal adoption time framework capturing the impact of rank, stock and order effects on timing of adoption. Assuming that firm i in industry j adopts a new technology by purchasing one unit of the technology at price P_t at time t. A function $g_i(\tau)$ defines the real benefits accruing to firm i in period τ from adopting the new technology. The real benefits of adoption at period τ can be represented by

$$g_{ij}(\tau) = f(R_i, S_{jt}, O_j) \qquad \tau \geq t, \ f_2 < 0, f_3 < 0 \qquad (4\text{-}1)$$

where R_i is the vector of the variable representing firm's heterogeneity and its inherent characteristics (i.e, rank effect), S_{jt} is the vector of the variable representing stock effects, which is the number of farmers already using the technology at time t, while the number of firms expected to adopt after t (O_j), captures the order effect. The order effect describes a firm's adoption decision which takes into consideration how waiting time affects its profits. For any given cost of acquisition, it will be profitable only for firms in certain order of adoption to actually adopt. The cost of acquisition is assumed to fall over time, and as it does so the number of adopters increases while profit declines. This maps out the diffusion path (Karshenas and Stoneman, 1993).

Defining r as the discount rate and assuming no depreciation of technological capital, the net present value of real benefits from adopting the technology at time t can be specified as

$$G_i(t) = -P_t + \int_t^\infty g(.) \exp\{-r(\tau - t)\} d\tau \qquad (4\text{-}2)$$

where P_t is the price of acquiring the technology. The choice of an optimal time (t^*) to adopt is determined by two conditions; the profitability or necessary condition, and the arbitrage or sufficient condition. The necessary condition requires that if adoption will occur, technology must yield positive profits relative to the use of traditional technology, i. e., $G_i(t) \geq 0$. It is the arbitrage condition that actually governs optimal adoption time, t^* for each potential adopter, and this is satisfied if net-benefit is not increasing over time, i. e.,

$$y_i(t) = \frac{d[Gi(t).\exp(-rt)]}{dt} \leq 0 \qquad (4\text{-}3)$$

We may then specify the optimal adoption date for firm i, as

$$y_i(t^*_i) \leq 0, \qquad (4\text{-}4)$$

Given that the distribution of the disturbance term μ remains invariant across firms over time, the stochastic component of the equation above can be specified as

$$y_i(t) + \mu \leq 0 \qquad (4\text{-}5)$$

The probability of adoption or hazard rate for a firm that has not adopted the technology at time t, given that it is at risk, can be expressed as

$$\lambda_i\,(t/X_i) = Prob[y_i(t) + \mu \leq 0] = V[-y_i(t)] \qquad (4\text{-}6)$$

where X_i represents farm and non-farm factors influencing adoption and diffusion of NERICA technology and λ_i is the probability or hazard of adoption. The probability of adoption or hazard rate of adoption is estimated using the duration model.

4.1.1 Duration analysis

Suppose a subject is being observed under a continuous time situation, the length of its spell is the continuous random variable τ with cumulative distribution function (cdf), $F(t)$; and probability density function (pdf), $f(t)$. $F(t)$ is also known as the failure function.

The survival function is given by

$$\begin{aligned} S(t) &\equiv 1 - F(t) \equiv Prob(\tau > t) \\ &\equiv -\exp\left(-\int_0^t \theta(s)ds\right) \end{aligned} \qquad (4\text{-}7)$$

where τ is length of a spell or the period for which an event is observed. Assuming that the entry time is known, t is the exact time at which the event actually occurs. The failure function, which is also the *cdf* is given by

$$F(t) = Prob\ (\tau \leq t) \qquad (4\text{-}8)$$

The *pdf* which is the slope of the *cdf* (i.e the failure function) which is given by

$$f(t) = \lim_{\Delta t \to 0} \frac{Prob\ (t \leq \tau \leq t + \Delta t)}{\Delta t} = \frac{dF(t)}{dt} = -\frac{dS(t)}{dt}$$

where t is the elapsed time between the beginning of a spell and its end, Δt is the very small (infinitesimal) interval of time. The $f(t)\Delta t$ is akin to unconditional probability of having a spell of length exactly t, i.e., leaving the state in tiny interval of time $[t, t+\Delta t]$. The continuous time hazard rate $\theta(t)$ is defined as

$$\theta(t) = \frac{f(t)}{1 - F(t)} = -\frac{f(t)}{S(t)} \qquad (4\text{-}9)$$

However, when the survival times are interval censored, it means that many events may occur within an interval (say a year) and they are all recorded as yearly integers. Suppose that the time axis is partitioned into a number of contiguous non-overlapping ('disjoint') intervals where interval boundaries are $t1, t2, t3, ..., tk$. (where the intervals need not be of equal length) and the spell of subject i in interval j begins at $t\text{-}1$ and ends at t, the value of the survival function at the time demarcating the start of the jth interval is

$$S(t\text{-}1) = Prob\ (\tau > t\text{-}1) = 1 - F(t\text{-}1) \qquad (4\text{-}10)$$

Where $F(.)$ is the failure function defined earlier. The value of the survival function at the end of the *jth* interval is

$$S(t) = Prob\ (\tau > t) = 1 - F(t) \qquad (4\text{-}11)$$

The probability of the exit within the interval *jth* (year) is given by

$$Prob(t\text{-}1 < \tau \leq t) = F(t) - F(t\text{-}1) = S(t\text{-}1) - S(t) \qquad (4\text{-}12)$$

Unlike the continuous time duration models in which the exact survival times of events are known with certainty, the underlying assumption of discrete time duration model is that failure events occur within an interval.

Therefore interval hazard rate $\lambda(t)$ also known as the discrete-time hazard rate is the probability of exit in the interval $(t\text{-}1, t)$. It is given by

$$\lambda(t) = Prob\ [\tau = t\ |\ t \geq t\text{-}1] = Prob\ [t-1 < \tau \leq t\ |\ \tau > t\text{-}1]$$

$$= \frac{Prob(t-1 < \tau \leq t)}{Prob\ (\tau > t-1)} \qquad (4\text{-}13)$$

where τ is the period for which a subject is being observed and t is the exact failure time, $\lambda(t)$ is the probability that failure event will occur at time t, given that it had not occurred at time $t\text{-}1$. Interval time hazard rate $\lambda(t)$ is a conditional probability that an event occurs at time t, given that it has not already occurred, such that $0 \leq \lambda(t) \leq 1$, whereas continuous time hazard is the instantaneous probability that an event occurs at time t, given that it has not already occurred. It is not really a probability since it may be greater than 1 (Allison, 1982).

Censoring and truncation are occurrences in which the survival times of a subject are not fully observed due to lack of complete information. While left censoring refers to a situation in which the beginning of an event is not known, right censoring refers to a situation in which the end time of an event is not known. On the other hand, left trucation refers to a situation where a subject enters late into a study, while right truncation refers to an early exit – for example, when a sample is drawn from the persons who exit from the state at a particular time. However in this study, if a farmer has not adopted the technology by the date of the survey, they are right-censored at the end of the observation period.

A major advantage of the data collected for this study is that cases of left censoring and right truncation did not occur. The likelihood function of the hazard rate for censored and uncensored observations in the sampled population is given by

$$\mathcal{L}i = \prod_{i=1}^{n} \left(\frac{\lambda_{it}}{1-\lambda_{it}}\right)^{ci} \cdot \prod_{k=1}^{t} (1-\lambda_{ki})$$

(4-14)

where c_i is a censoring indicator defined as $c_i=1$ if a spell is complete, and $c_i=0$ otherwise; k_i is a positive integer of the exact year of adoption. There are reported cases in which farmers entered into rice farming after the technology had been introduced, which is a typical case of left truncation however, no cases of right truncation is reported in the study. With delay entry at time (d_i) by farmer i, likelihood function above is conditioned on survival up to time (d_i). By taking left truncation into account, the likelihood function for censored and uncensored samples after is therefore as stated below (Jenkins, 2008).

$$\mathcal{L}i = \prod_{i=1}^{n} \left(\frac{\lambda_{it}}{1-\lambda_{it}}\right)^{ci} \cdot \prod_{k=di+1}^{t} (1 - \lambda_{ki})$$

(4-15)

4.1.2 The empirical models for adoption and diffusion of NERICA technology

Two main specifications of discrete-time hazard models exist in the literature; they are the proportional hazard and proportional odd models. Proportional hazard (PH) model relates the time that passes before an event occurs to the covariates that may be associated with that quantity of time. In PH models, a unit increase in a covariate has a multiplicative effect on the hazard rate. On the other hand, the proportional odds models are useful for fitting data whose hazard rates converge asymptotically. The regression parameter estimates are interpreted as the additive change in the log-odds of survival associated with a one-unit change in covariate values (Jenkins, 1995; Rossini and Tsiatis, 1996). The Akaike and Bayesian information criteria were employed to decide on which of the two models is appropriate for this study.

The results (reported in chapter 6) show that proportional hazard is preferred over its odds counterpart. The general form of the discrete-time proportional hazard model is given by

$$\lambda(t, X_{it}, \beta) = \lambda_0(t) \exp(X_{it}, \beta) \qquad (4\text{-}16)$$

The discrete time counterpart of proportional hazard model is given by

$$\lambda(t, X_{it}) = 1 - exp[- exp\ (\beta'X_{it} + \lambda_0)] \qquad (4\text{-}17)$$

where λ_0 is the baseline hazard rate, X is a vector of variables the determine farmers' optimal choice and β is a vector of parameters to be estimated. The present study employs Weibull distribution for the baseline hazard. The Weibull distribution is particularly suitable for modeling data that exhibit monotone hazard rates – increasing or decreasing trends of events over time, which is a typical case of agricultural diffusion pattern. The empirical diffusion path in figure 6-1 suggests that the rate of diffusion is not uniform over time. This indicates that the probability that a farmer will adopt the technology, given that he or she had not previously adopted appears to be increasing over time. Hence, the assumption of a constant hazard function seems to be unreasonable (Heckman and Singer, 1984).

The dependent variable is a dichotomous variable, which is a combination of the censoring status of the farmer and the year of technology adoption. To create the binary dependent variable for each subject, the data are organized such that farmer i has n number of rows corresponding to the number of years he is at risk of experiencing failure (maximum of 12 per farmer in our case), resulting in multiple rows of unbalanced panel data. If farmer i's survival time is not censored, the binary dependent variable is equal to 0 for all the years he is at risk, but equal to 1 for the year he adopted the technology. Thus, subjects were observed over a total period of 3731 times for which they were at risk. We then proceeded to estimate the model as complementary log-log regression model.

A potentially important issue in duration models is unobservable characteristics. For example, there could be some unobserved characteristics of a farmer's management style that favors adoption of the technology, or learning by doing when cultivating a new variety. Thus, if unobserved heterogeneity is present, but not accounted for, this will tend to bias the coefficient of any variable with which it is correlated. In particular, the baseline hazard will pick up unobserved firm-specific heterogeneity, resulting in a downward bias in the degree of duration dependence (Lancaster, 1990). Therefore, a duration model that incorporates a term for unobserved farm household heterogeneity is employed in this study.

4.2 Conceptual framework for determinants and impact of technology adoption

Following Abdulai and Huffman (2014), farm households are assumed to self-select themselves into adoption of NERICA technology or otherwise by considering the net benefits accruable from adopting the technology. Farm households therefore choose the option that provides maximum net benefits. Under this assumption, let us represent the net benefit that household j derives from adopting the technology as Y_{jA} and the net benefit from non-adoption represented as Y_{jN}, with net benefits representing wealth, then the two regimes can be specified as:

$$Y_{jA} = X_j\beta_A + \varepsilon_{jA}$$

and (4-18)

$$Y_{jN} = X_j\beta_N + \varepsilon_{jN}$$

where X_j is vector of the variables representing factor prices as well as farm and household characteristics, β_A and β_N are vectors of parameters and ε_{jA} and ε_{jN} are iids. Farm households will normally choose the technology if the net benefits obtained by doing so are higher than those obtained by not using the technology, that is, $Y_{jA} > Y_{jN}$. Thus, farmer adopts the technology only if the perceived net benefits are positive. Although the preferences of the farmer, such as perceived net benefits of adoption are unknown to the researcher, the characteristics of the farmer and the attributes of the technology are observed during the survey period. Thus, the net benefits derived from technology adoption can be represented by a latent variable D_j^*, which is not observed but can be expressed as a function of the observed characteristics and attributes denoted as Z in the latent variable model as follows:

$$D_j^* = Z'\gamma_j + \mu_j, \quad [D=1, \text{ if } D_j^* > 0] \qquad (4\text{-}19)$$

where D_j is a binary variable that equals 1 for farmers that adopt the technology and zero otherwise, with γ denoting a vector of parameter to be estimated. The error term μ_j with mean zero and variance δ^2_u captures measurement errors and factors unobserved to the researcher but known to the farmer. Variables in Z include factors influencing the adoption decision, such as farm-level and household characteristics therefore, equation 4-19 is also known as selection equation. The probability of technology adoption can then be expressed as

$$\Pr(D_j=1) = \Pr(D_j^* > 0) = \Pr(\mu_i > Z'\gamma) = 1 - F(-Z'\gamma) \qquad (4\text{-}20)$$

4.2.1 Empirical models for determinants and impact of technology adoption

As indicated above, the two regimes representing the net benefits derived from technology adoption by adopting and non-adopting households are given by

Regime 0 (Non-adopters) $\quad Y_{jA} = X_j \beta_A + \varepsilon_{jA} \quad$ if $D_j = 0$

Regime 1 (Adopters) $\quad Y_{jN} = X_j \beta_N + \varepsilon_{jN} \quad$ if $D_j = 1$

(4-21)

where Y_{jA} and Y_{jN} are the net benefits (outcome of technology adoption) such as net-returns and poverty status of farm households in regimes 1 and 2. X_j represents vector of the exogenous variables thought to influence the outcome function, while β_A and β_N are parameters to be estimated and ε_{jA} and ε_{jN} are error terms. However, selection bias may occur if unobservable factors influence the error terms in the section equation (4-19) and the outcome equations (4-21) thus, resulting in correlation between the two error terms such that $corr(\varepsilon, \mu) = \rho \neq 0$.

In order to account for selection bias that may arise from observable and unobservable in farm and non-farm characteristics of the farm households on one other hand and estimate impact of NERICA technology adoption on the outcomes of interest on the other hand, the Endogenous Switching Regression (ESR) model approach (Lee, 1982; Maddala, 1986) is employed. In ESR model, the error terms of the selection and outcome equations are assumed to have a trivariate normal distribution, with zero mean and non-singular covariance matrix expressed as:

$$cov(\mu_i, \varepsilon_1, \varepsilon_2) = \begin{bmatrix} \sigma_A^2 & \sigma_{AN} & \sigma_{A\mu} \\ \sigma_{AN} & \sigma_N^2 & \sigma_{N\mu} \\ \sigma_{A\mu} & \sigma_{N\mu} & \sigma^2 \end{bmatrix} \quad (4\text{-}22)$$

$\sigma_A^2 = var(\varepsilon_1)$; $\sigma_N^2 = var(\varepsilon_2)$ ($\sigma^2 = var(\mu_i)$; $\sigma_{AN} = cov(\varepsilon_1, \varepsilon_2)$; $\sigma_{A\mu} = cov(\varepsilon_1, \mu_i)$; and $\sigma_{N\mu} = cov(\varepsilon_2, \mu_i)$; σ^2 represents the variance of the error term in the selection equation and σ_1^2, σ_2^2 represent the variance of the error terms in the outcome equations.

According to Johnson and Kotz (1970), the expected values of the truncated error terms are given by

$$E(\varepsilon_{JA}|D_i=1) = E(\varepsilon_{JA}|\mu_i > -Z'\gamma_j) = \sigma_{A\mu}\left[\frac{\varphi(Z'\gamma_j/\sigma)}{\Phi(Z'\gamma_j/\sigma)}\right] \equiv \sigma_{A\mu}\lambda_1$$

(4-23)

$$E(\varepsilon_{JN}|D_i=0) = E(\varepsilon_{JN}|\mu_i \leq -Z'\gamma_j) = \sigma_{N\mu}\left[\frac{-\varphi(Z'\gamma_j/\sigma)}{1-\Phi(Z'\gamma_j/\sigma)}\right] \equiv \sigma_{N\mu}\lambda_2$$

where φ and Φ are the probability density and cumulative distribution functions of the standard normal distribution respectively. The ratio of φ and Φ represented by λ_1 and λ_2 in equations (4-23) is referred to as the Inverse Mills Ratio (IMR) which denotes selection bias terms. Equations in (4-21) can then be written as

Adopters: $\quad Y_{JA} = Z'\beta_{JA} + \sigma_{A\mu}\lambda_1 + \varepsilon_{JA}$

(4-24)

Non-Adopters $\quad Y_{JN} = Z'\beta_{JN} + \sigma_{N\mu}\lambda_2 + \varepsilon_{JN}$

While previous studies have employed a two-stage method to estimate this parametric procedure by deriving inverse Mills ratios λ_1 and λ_2 from the selection equation in the first stage and subsequently incorporating these Mills ratios into the second stage estimation (Freeman et al. 1998; Abdulai and Binder, 2006), this two-step procedure may generate heteroskedastic residuals that cannot be used to derive consistent standard errors without cumbersome adjustments (Maddala, 1986). Therefore, the present study employs the single stage Full-Information Maximum Likelihood (FIML) method proposed by Lokshin and Sajaia (2004). The FIML method fits the selection and outcomes equations simultaneously in order to yield consistent standard errors thus, λ_1 and λ_2 in equations (4-24) are homoskedastic.

The FIML's log likelihood function for switching regression model employed in this study proposed by Lokshin and Sajaia (2004) is described below:

$$\ln D_i = \sum_{i=1}^{N} \left\{ \begin{array}{l} D_i w_i \left[\ln F(\dfrac{Z'\gamma_j + \rho_{1\mu}(Y_{jA} - X_j\beta_A/\gamma_j)}{\sqrt{1-\rho_{A\mu}^2}}) + \ln(f((Y_{jA} - X_j\beta_A/\gamma_j)) \right] \\ +(1-D_i)w_i \left[\begin{array}{l} \ln(1 - F(Z'\gamma_j + \rho_{2\mu}(Y_{jN} - X_j\beta_N)/\gamma_j) \\ \sqrt{1-\rho_{N\mu}^2} \\ +\ln(f((Y_{jN} - X_j\beta_N)/\gamma_j) \end{array} \right] \end{array} \right\} \quad (4\text{-}25)$$

There is endogenous switching if either $\rho_{A\mu}^2$ or $\rho_{N\mu}^2$ is statistically significant. Specifically, if $\rho < 0$, this would imply negative selection bias, indicating that households with below average net-returns are more likely to adopt the technology. On the other hand if $\rho > 0$, it implies positive selection bias suggesting that farmers with above average net-returns and household welfare are likely to adopt the technology.

The expected outcomes of the adopting and non-adopting households are stated as follow:

$$E\left(Y_{jA} | D_i = 1\right) = Z'\beta_{jA} - \sigma_{A\mu}\lambda_1$$
$$E\left(Y_{jN} | D_i = 1\right) = Z'\beta_{jN} - \sigma_{N\mu}\lambda_2$$
(4-26)

The average treatment effect on the treated (*ATT*) of NERICA technology adoption can be calculated as

$$ATT = E(Y_{JA} - Y_{JN}|D_i = 1) = Z'_i(\beta_A - \beta_N) + (\sigma_{A\mu} - \sigma_{N\mu})\lambda_1 \tag{4-27}$$

4.3 Conceptual framework for smallholder market participation

According to Strauss (1984) and Goetz (1992), marketed surplus is the difference between household supply (production) and demand (consumption) functions. If household supply function is represented by

$$x_j = x_j(p, \varphi) \tag{4-28}$$

and consumption function is given by

$$x_j^c = x_j^c[p, \eta, \alpha + p_n T(\mu) + f(p, \varphi)] \tag{4-29}$$

then, absolute value of the marketed surplus *(q)* of commodity *j* is calculated as

$$|q_j| = x_j(.) - x_j^c(.) \tag{4-30}$$

where x_j refers to production and x^c_j to consumption of good *j*; *p* is a vector of prices; p_n is price of labor; η denotes household characteristics affecting taste; α is exogenous income; *T* is time available for work and leisure; μ is household characteristics determining *T*; φ is a vector of farm characteristics including fixed inputs and a vector of production technology parameters, and *f* is profits.

Assuming that farm household produces two goods j and k, then, marketed surplus elasticity of good *j* with respect to price of good *k*,

$$\frac{p_k}{|q_j|} \cdot \frac{\partial q_j}{\partial p_k} = \frac{x_j}{|q_j|} \frac{p_k}{x_j} \frac{\partial x_j}{\partial p_k} - \frac{x_j^c}{|q_j|} \frac{p_k}{x_j^c} \frac{\partial x_j^c}{\partial p_k} \tag{4-31}$$

which represents the difference between household supply (production) and demand (consumption) elasticities, weighted respectively by the ratios of quantities supplied and demanded, hence the absolute value of the marketed surplus. Note that for households not participating in the market for commodity *j*, the elasticity is undefined. Assuming that production and consumption decisions are made simultaneously, Strauss (1984) derived the reduced form of marketed surplus as

$$q_i = q_i(p, \eta, \alpha, \mu, \varphi) \equiv q_i(x_i^q) \tag{4-32}$$

Given that farm income Y_i accruable to randomly selected rice producing household depends on the volume of marketed surplus q_i, the discrete choice variable representing market participation decision is Di, and farm household's observable characteristics are Xi; then, we can express Y_i as a linear function of X_i and D_i respectively as follows:

$$Y_i = \alpha X_i + \delta D i + \varepsilon_i, \tag{4-33}$$

where α and δ are vectors of parameters to be estimated, and ε is the error term. The impact of market participation on the outcome variable Y_i is measured by estimate of the parameter δ. However, farm households are not randomly assigned into market participation or otherwise, thus, variable D_i is not exogenous and OLS estimation of equation (4-33) may be biased.

In order words, market participation decision D_i can be explicitly expressed as

$$D_i = Z'\gamma_i + \mu_i \tag{4-34}$$

Selection bias occurs if there is correlation between the error terms of the outcome equation (4-33) and that of the selection equation (4-34). Lee (1982) developed the endogenous switching regression model as a generalization of Heckman's selection correction approach. In the switching-regression approach, two equations rather than one with a dummy variable representing market participation status allows coefficients to differ between market participation regimes. Thus, market participation is allowed not only to have an intercept effect on the outcome, but also to exert slope effects that may be different between the two groups.

The rationale behind Heckman selection estimator is to control directly for the part of the error term in the outcome equation that is correlated with the selection equation dummy variable. Equation (4-35) expresses the relationship between farm income and the explanatory variables:

Regime 1: Participants $Y_{i1} = \gamma X_{i1} + \delta D_{i1} + \varepsilon_1,$

(4-35)

Regime 2: Non-participants $Y_{i2} = \gamma X_{i2} + \delta D_{i2} + \varepsilon_2,$

4.3.1 Empirical models for determinants and impact of market participation

The empirical estimations of determinants of market participation and its impact on economic returns and welfare of rice producing households follow the endogenous switching regression empirical strategy described in section 4.2.1 above. As indicated previously, the outcome variables of interests are Return on Investments (ROI) and poverty incidence. ROI is an indicator that takes into account the fact that farmers operating as entrepreneurs do not only concentrate on improving farm income, but also consider the profitability of their investments (Asfaw et al. 2009). The approach is a widely used relative profitability performance measure of management control for a single investment. ROI is expressed as

$$ROI = \frac{Profit}{Investment} \tag{4-36}$$

where investment in this case involves all cost incurred during production, processing and marketing of rice in 2011.

The advantage of ROI compared to other measures such as net income is that it relates profit to farmer's investment decision and consequently indicates how well the available assets have been used (Kleemann et al. 2014). The two main measures of household welfare used in this study are poverty head-count and poverty gap. These measures were computed using household incomes for the year 2011. Since there is no country level poverty line for Nigeria (i.e a threshold below which a given household or individual can be classified as poor), household poverty was computed using the World Bank's threshold of $1.25 (at 2005 purchasing-power parity PPP) minimum level of income per person per day.

Poverty headcount index measures the proportion of the population whose per capita income falls below the poverty line and hence measures the incidence of poverty. However, it does not take intensity of poverty into account. Poverty gap index is an improvement over the poverty headcount index in that the depth of poverty reflecting how far the poor are from the poverty line is measured. This measure is also regarded as the cost of eliminating poverty as it shows how much would have to be transferred to the poor to bring their incomes up to the poverty line.

Poverty severity index is basically the square of the poverty gap. This is simply a weighted sum of poverty gaps (as a proportion of the poverty line), where the weights are the proportionate poverty gaps themselves (Coudouel et al. 2002). The empirical estimation of poverty indices follows the Foster–Greer–Thorbecke (FGT) (1984). The FGT poverty measure can be expressed as

$$P_x = \frac{1}{N} \sum_{i=1}^{n} \left[\frac{z - yi}{z} \right]^x \qquad (4\text{-}37)$$

where N is the number of people in the sample population, z is the poverty line, y is per capita income for the *ith* household, and x is the poverty aversion parameter. When $x = 0$, P_x is simply the headcount index or the proportion of people that is poor. When $x = 1$, P_x is the poverty gap index, a measure of the depth of poverty defined by the mean distance to the poverty line, where the mean is formed over the entire population with the non-poor counted as having a zero poverty gap. When $x = 2$, P_x is a measure of severity of poverty and reflect the degree of inequality among the poor.

Chapter Five

Household Survey and Data Collection

Introduction:

This chapter provides information on the household level survey conducted to collect data on farm and non-farm characteristics of rice producers in Nigeria. A brief description of the study area is provided in section 5.1. Section 5.2 gives detailed account of the process of data collection, while an overview of the descriptive statistics of the respondents is provided in section 5.3. Farm household's perception of the constraints to rice production and marketing is provided in section 5.4.

5.1 The Study Area

Rice can be produced in almost all parts of Nigeria and in different agro-ecological zones; its cultivation is widespread under five major production systems classified as rain-fed upland, rain-fed lowland, irrigated, deep water and mangrove swamp (Akpokodje et al. 2001). The NERICA technology are upland rice varieties, data used for this study were collected from rice producing households in Ogun and Ekiti States, which are the major hubs of upland rice production in Nigeria. The two States are located in the Southwest region which comprises four other States, namely; Oyo, Osun, Ondo, Ogun and Lagos. The Southwest region lies between longitude 2^031^1 and 6^000^1 East and latitude 6^021^1 and 8^037^1 North with a total land area of 77,818 km^2 and an estimated population of about 30 million people. The region is predominantly inhabited by the *Yorubas* who mainly live in the rural areas and have farming as their major occupation. Rice is a major food crop grown in the two States and the crop is produced for sale and for home consumption. Figure 5-1 shows the study area

Maps

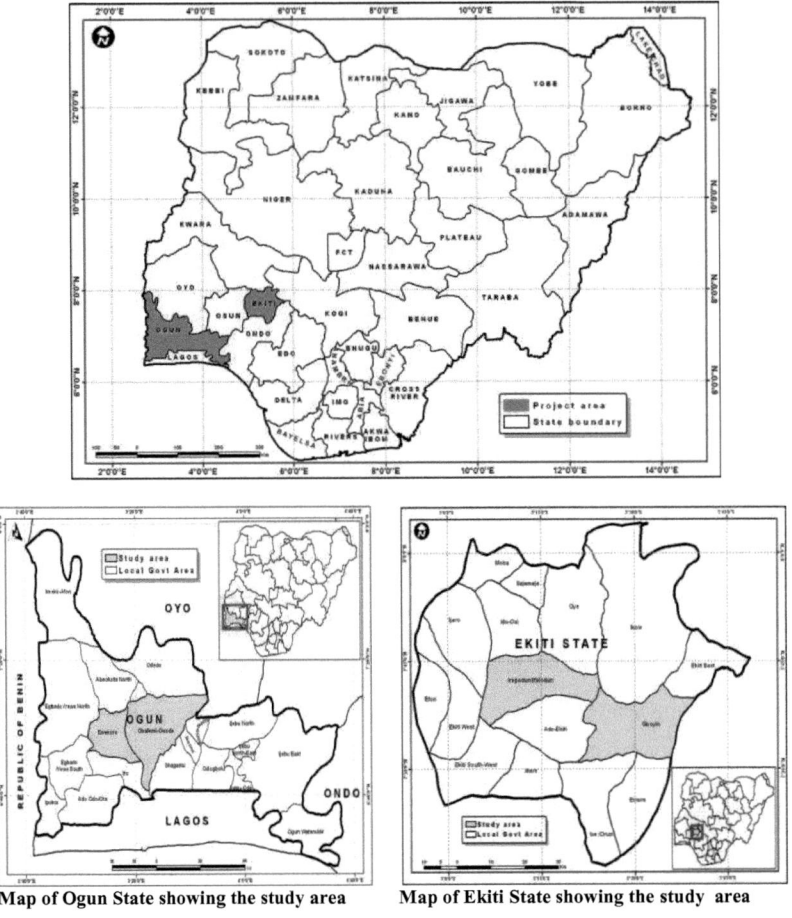

Map of Ogun State showing the study area   Map of Ekiti State showing the study area

Figure 5-1: Maps showing the study area

5.2 Sampling procedure and data collection

The data used for this study were collected from rice producing households in Ekiti and Ogun States between May and August, 2012 using multistage sampling procedure. In the first stage, the two States were purposively selected due to their prominent positions as the major hubs of upland rice production, processing and marketing in Nigeria. A series of meetings were held with relevant stakeholders in the rice sector in each of these two States to collect firsthand information on rice production in general and NERICA dissemination. Amongst the stakeholders met during the course of the survey were lead farmers, extension personnel and producers' cooperatives.

Focus group discussions were also held in 2 locations per State in order to pretest the questionnaire and gather general information on rice production, processing and marketing at State level. The second stage of the sampling procedure involved selection of two local government areas in each State fairly reflecting the major agricultural ecologies and different levels of private and public agricultural technology supports based on the information made available by the local extension personnel. The local government areas were then stratified into villages with high and low concentrations of agricultural innovation and marketing activities, while 23 representative rice producing communities were randomly selected. Finally, 400 households were randomly selected in proportion to the population of rice producing households in these communities.

However, only 380 questionnaires contained complete information and were therefore used for the study. Information was collected on general socioeconomic, farm and non-farm factors as well as innovation behaviour of farm households with the aid of a pretested questionnaire and assistance of trained enumerators. Information gathered on production activities, marketing activities and farm household's income was limited to January – December 2011 for consistency. With respect to technology adoption, information was gathered on the history of NERICA adoption from 2001 when the technology was introduced to the year of the survey.

5.3 An overview of the descriptive statistics of the farm households surveyed

A general overview of the socioeconomic characteristics of the respondents is presented in table 5-1. Ordinarily, Nigerian smallholder farmers are ageing and are quite old. As noted by Farinde et al. (2007), most of the young able-bodied men and women have migrated from the rural to the urban centres in search of better lives and the older generation is now left on the farm. However, rice farmers are of middle age with a mean of about 44.2 years. This can be attributed to the rigours of rice production which may be difficult for old and aging people to cope with. About 89 percent of the household heads are men. This shows that rice farming is basically men's affairs. Although rice production tasks are generally allocated along gender lines, rice production in the study area is basically classified as men's tasks due to the drudgery involved. Nevertheless, women help out with tasks such as transplanting of seedlings to the fields, harvesting and threshing (ODI, 2000). Farm size measured in hectares is the land area used by farmers for rice cultivation during 2011 cropping session. The areas of land cultivated range from 0.2 to 6 hectares with a mean of 1.577.

Table 5-1 Descriptive statistics of the farm household surveyed

Variables Name	Description	Mean	Std. Dev.	Min.	Max.
Age	Age of farmer (Years).	44.189	8.247	20	85
Farm size	Size of rice farm cultivated in 2011 (Hectares).	1.576	1.021	0.2	6.0
Education	Education of the household head (Years).	9.045	4.143	0	18
Extension distance	Distance to extension office (Km).	21.544	4.457	8	35
Access to extension	1 if farmer had contact with extension agent, 0 otherwise.	0.452	0.498	0	1
Accesses to credit	1 if household head is not liquidity constrained, 0 otherwise.	0.266	0.442	0	1
Gender	1 if household head is male, 0 otherwise.	0.889	0.3.3	0	1
Village market	1 if there is market in farm household's community, 0 otherwise.	0.542	0.499	0	1
Market distance	Distance of the farm household to market (Km).	2.609	3.205	0	15
Household size	Number of the people living together in the same house and eating from one pot.	5.244	1.994	1	12
Land ownership	1 if the plot of land where rice is cultivated is owned by the household, 0 otherwise.	0.479	0.300	0	1
Adoption status	1 if farm household adopts NERICA technology, 0 otherwise.	0.489	0.510	0	1
Market participation status	1 if farm household participates in markets, 0 otherwise.	0.576	0.495	0	1
Electricity	1 if household has electricity connection, 0 otherwise.	0.723	0.447	0	1
Ownership of livestock	1 if household rears livestock, 0 otherwise.	0.363	0.482	0	1
Fertilizer application	1 if household uses fertilizer, 0 otherwise.	0.292	0.455	0	1
Herbicides application	1 if household uses herbicides, 0 otherwise.	0.405	0.492	0	1
Group membership	1 if household belongs to farmers' association, 0 otherwise.	0.508	0.501	0	1
Distance of the community to tarred road	Distance from farm household community to the nearest tarred road measured in (Km).	2.545	2.69	0.03	14
Presence of primary heath care centre in the community	1 if heath care centre is present in the community, 0 otherwise.	0.589	0.493	0	1
Possession of radio	1 if household owns a radio set, 0 otherwise.	0.776	0.567	0	1

Possession of mobile phone	1 if household possess a mobile phone, 0 otherwise.	0.479	0.510	0	1
Off-farm income	1 if household earns off-farm income, 0 otherwise.	0.395	0.489	0	1
Net-returns (Naira)	Revenue minus expenditure on variable inputs.	280298.1	266727.3	0.01	1699733
Poverty headcount	1 if household is poor, 0 otherwise.	0.639	0.480	0	1
Poverty gap	Measures the distance of household from the poverty line.	0.316	0.268	0	0.980
Poverty severity	Measures the square of the poverty gap.	0.199	0.258	0	0.950
Labour constraint	1 if farmer is labour constrained, 0 otherwise.	0.839	0.368	0	1

Source: Field Survey, 2012.
Exchange rate in 2012, $1 = 160 Naira

Given that rice is the major crop grown in the study area and source of household income, farm size are classified into 4 categories to quantify volume of farming activities of farm households. Majority of the farmers interviewed cultivated between 1 and 2 hectares (see figure 5-2). This is consistent with smallholder land holding not only in Nigeria but across the sub-Saharan African.

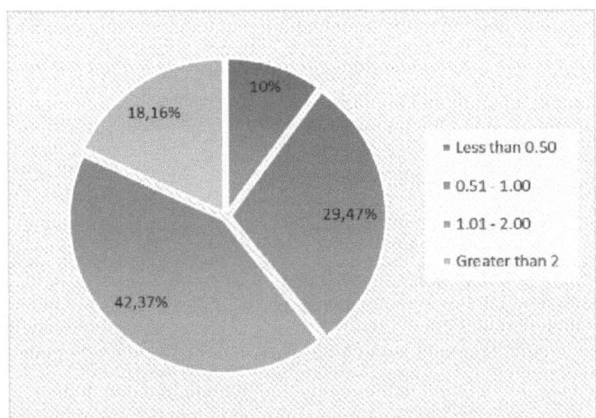

Figure 5-2: Distribution of area of land under rice cultivation by farm households.

Farmers in the study area are quite educated. The average years of education of the household head is found to be about 9 years. This implies that the majority of the rice producers have at least primary education. Household size is defined as the number of people living together and eating from the same pot. The average household size of rice producing households in the study area was found to be about 5 people per household. With respect to assets ownership, about 48% of the farm households owns mobile phone, 77% has radio set, and 39.5% earns off-farm income while about 37% owns livestock. About 54% of the households surveyed indicated that they have access to markets where inputs and farm produce are treaded in their communities, while average distance to market measured in Kilometers is found to be 2.61.

Information was also collected on distance of the community to tarred roads as this variable may be important in explaining access of farm households to markets, extension services and other infrastructure. Average distance to tarred roads is found to be 2.55km. Primary health care facilities are present in almost every community and it is located in about half of a Kilometer from farm households' residence. Although electricity is in a deplorable state in Nigeria, about 73% of the respondents has electricity connection in their homes.

5.4 Constraints to rice production and marketing

A major constraint to rice production in Nigeria is access to land. Although land is relatively abundant in the country, there are limitations to gaining access to land for agricultural purposes. The prevalent land tenure system especially in the rural area is such that land is usually owned by traditional extended family system and it can mostly by acquired through inheritance. Land acquisition through inheritance tends to perpetuate land fragmentation leading to reduced land/person ratio, shortened fallow periods, reduced soil productivity and increasing environmental problems (Etim and Edet, 2013).

In the present study, only 48% of the farm households sampled produced rice on farm land owned by them, while the remaining 52 percent are tenants. Apart from land constraints, other constraints facing the rice producers include labour, credit, marketing constraints, as well as low access to agricultural extension services. For instance, Ebukiba (2010) noted that labour accounts for the highest share of the cost of agricultural production in Nigeria due to its continuous movement to the other sectors of the economy. In some countries, the constraints imposed by inelastic supply of labour have been successfully offset by substitution with mechanized power however, mechanization is not within the reach of most of the respondents. Figure 5-3 presents information on the constraints faced by rice producers in Nigeria.

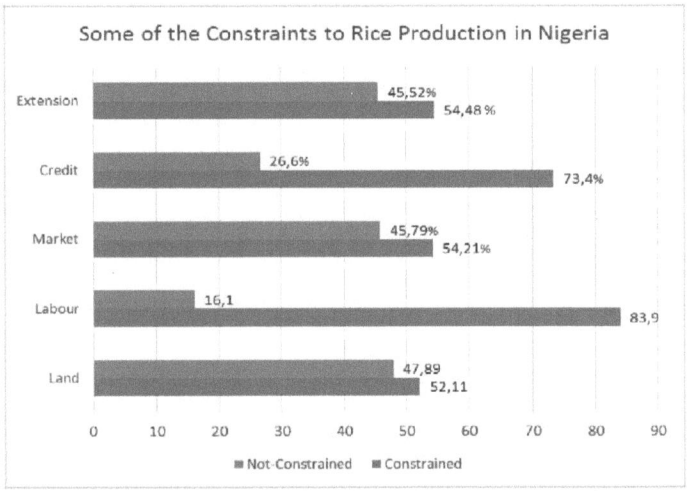

Figure 5-3: Constraints to rice production in Nigeria
Source: Field survey 2011

Chapter Six

Empirical Results

Introduction:

This chapter is divided in three main sections. Section 6.1 presents the empirical results of adoption and diffusion of NERICA technology. Section 6.2 presents the empirical results of impact of NERICA adoption on net-returns and welfare of rice producing households, while section 6.3 shows the empirical results of impact of market participation on returns on investment (ROI) and welfare of rice producing households.

6.1 Adoption and Diffusion of NERICA Technology

For the purpose of this study, adopters are defined as farmers who cultivated NERICA varieties fully or partially in addition to the traditional rice varieties on their farms. Adoption spells began in the year 2001 when the technology was made available to rice producers in the study area and ended whenever individual farmer adopted the technology. As at the time of the survey, only 186 out of the 380 farmers had complete adoption spells. Although adoption rate in the study area was initially slow, it became significantly higher from the year 2006 when farmers were fully aware of the technology.

However, the spells of the farmers who were yet to adopt the technology were right censored at the end of the observation period, showing that the process is ongoing and might occur in future. In cases of delayed entry, dates of entrance into rice farming were used as the beginning of the spell. Our estimations took into account the right censored and left truncation nature of the data collected. Figure 6-1 depicts diffusion of NERICA technology amongst the producers in the sampled population.

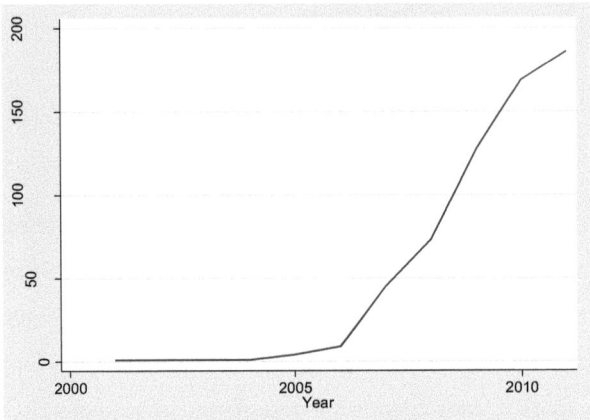

Figure 6-1: Adoption of New Rice for Africa in Nigeria

6.1.1 Variables included in the model

The explanatory variables are divided into time-invariant and time-varying variables. Time-invariant covariates are variables that are static over time (i.e, farm size, education, access to information, gender, market access, membership of FBG and farmer-neighbors interaction), while time varying variables include price of technology, age of farmer and time variable indicating the duration dependence of the baseline hazard. Firm size is a key variable in rank effect, this is the land area (in hectares) used by farmer for rice cultivation. Karshenas and Stoneman (1993) argued that size has a net positive effect on harzard of adoption due to economies of scale resulting in adoption of new technologies at lower costs. Moreover, larger farms are willing to undertake more risk than smaller farms, and as such are more likely to adopt. However, as noted by Zilberman (1984), larger farms may have less financial pressure to search for new ways of improving productivity and incomes, thus making the effect of farm size on adoption time ambiguous.

Access to production information and extension services are measured by the distance of the farm household to the nearest extension office as well as a dummy variable to capture farmer's access to extension services. Farmers who are farther away from extension office are likely to have lesser access to extension services and this may have negative effect on the tendency to adopt the technology. Education is measured by number of years of schooling by farmers. Farmers with more education are expected to have higher abilities to evaluate a technology and possibly adopt faster (Huffman, 2001).

Age of the household head is sometimes used as proxy for farming experience and this may influence adoption of a new technology. However, literature is divided over the expectation of the effect age of farmer on technology adoption. For example, Dadi et al., (2004) showed that age had negative effect on conditional probability of fertilizer adoption in Ethiopia, whereas Murage et al. (2011)'s findings on adoption of *Striga* weeds control technology was contrary. The expectation of farmer's age on adoption probability is therefore ambiguous.

Given that smallholder farmers are risk averse, production risks and uncertainty may deter them from adopting a new technology. Risk variables are computed from moments of farm level profit distributions following Koundouri (2006) and Genius et al. (2013). In the first step, the sample moments of the profit distribution for each farmer is computed. In a second step, the estimated moments of profits are included in the adoption and diffusion model. Specifically, the first stage involves regressing farm-level profit against farm inputs and farmers' socioeconomic characteristics, farm-level and village effects in the following specification;

$$\pi_{ij} = h(X_i, x_{ij}, Z_{ij}; \beta) + \varepsilon_i \qquad j = 1, 2 \qquad (6\text{-}1)$$

where π represents profits per acre, i denotes individual farmers, X_i is a vector of variable inputs used per acre by farmer i, x_{ij} denotes specific inputs related to the technology j, such as labor for farmer i and Z is a vector of variables capturing farm-level and household level characteristics of the farmer, ε is a random variable with zero mean. Given the assumption of expected profit maximization, the explanatory variables in equation (6-1) are assumed to be predetermined. This implies that ordinary least squares (OLS) approach would produce consistent and efficient estimates of the parameters vector in equation (6-1) (Antle, 2010). The results of the first stage farm-level profit regression are presented in table 6-1 below.

Table 6-1: Ordinary least square estimates of the parameter of profit function

	Coefficient	S.E
Constant	-14.985***	1.400
Age	0.456***	0.048
Age squared	-0.005***	0.0005
Gender	2.207***	0.219
Education	-0.202***	0.065
Education squared	0.011****	0.004
Farm size	7.204***	0.225
Farm size squared	-1.073***	0.441
Market price	0.028***	0.002
Fertilizer cost	-0.009***	0.002
Seed cost	-0.004	0.003
Herbicide cost	-0.008***	0.0001
Labour cost	-0.005***	0.0003
Market distance	-0.112***	0.029
Access to extension	-0.152	0.199
Yield	3.410***	0.198
Access to extension X Market distance	0.205***	0.045
Access to extension X Market price	-0.005***	0.001
Obafemi-Owode	0.226	0.183
Ewekoro	-0.497**	0.211
Ifelodun-Irepodun	0.164	0.210
R^2	45.20	
Adjusted R^2	44.90	

*, **, and *** = 10%, 5% and 1% level of significance respectively

Generally, the kth central moment of profit ($k=2,3$), conditional on input use can be computed as

$$e^k = E\left\{\left[\pi(\cdot) - \varepsilon^1 \vert X_i, x_{ij}\right]^k\right\}, \qquad k = 2,3 \qquad (6\text{-}2)$$

where ε^1 is the first moment of profit. The higher central moments of profits are then specified as

$$\hat{\varepsilon}_{ij}^k = g(X_i, x_{ij}, Z_{ij}; \alpha) + \tilde{\varepsilon}_{ij} \qquad k = 2,3 \qquad (6\text{-}3)$$

Thus, the second central moment or variance of profit, π, is estimated by squaring the residuals and regressing on the same set of explanatory variables, while the third central moment (measuring skewness of profits) and fourth central moment (measuring kurtosis of profits) are obtained by utilizing the estimated errors raised to the third and fourth powers, respectively.

Farmers were asked to assess and rank the quality of the soil of their farms as either fertile or not fertile. Soil infertility may be a disincentive to technology adoption thus, the variable is expected to have negative effect on probability of technology adoption. Farmers who are not liquidity constrained are likely to adopt a new technology because they have resources to acquire the new technology and the supporting inputs.

Farmers are classified farmers as liquidity-constrained if they sought for, but were unable to obtain credit for their farm operations, while farmers who were able to obtained credit when needed are not liquidity-constraint. A dummy variable is created to capture membership of farmers' based group. In the same vein, it is a common practice in the study area for farmers to interact with fellow farmers on problems and prospects in their farm operations irrespective of their adoption status. Therefore, the respondents are classified into two broad groups based on whether or not they interacted with other farmers on their production activities. Specifically, farmers who interacted with their homophilc neighbors within half of a Kilometer radius to their homes or farmsteads are taken into account. Consequently, group membership of FBG and interaction with neighbors are expected to have positive influence on the hazard of adoption through learning effects.

The inherent characteristics of farmers (such as farm size, age, gender and education), which are termed rank effects in the diffusion literature, are expected to have statistically significant effects on the probability of adoption. Although the accumulated number of adopters to current date is expected to have negative effect on adoption probability, the cumulative numbers of previous adopters (S_{jt}) representing stock effect is expected to have a positive influence on adoption due to epidemic and learning effects. However, if the number of farmers expected to adopt the technology after time t representing order effect is positive and significant, it shows that rice producers in the study area are heterogeneous in order. The order effect is computed as the difference between cumulative number of adopters and previous adopters (i.e, S_{jt+1} - S_{jt}). In order to control for age as time varying covariate, age of farmer at the time of adoption is used for analysis. This is achieved by using episode splitting method. The price of purchase of the technology is the reported purchase price deflated by the consumer price index while the expected price change for the technology is measured by $P_{t+1} - P_t$.

The explanation variables in the duration model contain variables that are expected to either speed up or retard adoption of the technology. Such variables include age and education of farmer, liquidity constraints, visit by extension agents, and membership in farmers' organizations to build social capital, as well as district fixed effects. Farm-level variables include farm size, land quality, and distance of farmer's house to market. Of these variables, access to credit and contact with extension agents may be endogenous. For example, farmers with high potential to adopt the technology may have higher probabilities of being visited by extension agents. The decision to adopt the technology and access to credit may also be jointly determined, since farmers adopting the technology will have higher capital requirements to pay for hired labor. Given the discrete nature of the dependent variable, we employ the Blundell and Smith (1989) approach to account for the potential endogeneity of these variables.

The estimation is carried out by first specifying the potential endogenous variables as functions of all other explanatory variables in the hazard model in addition to a set of instruments in the first-stage regressions. Rather than using the predicted values from the first-stage equation as in a commonly used two-stage estimation approach, Blundell and Smith (1989) approach involves specifying the hazard model and then including the observed values of the endogenous variables as well as the residual terms from the first-stage regressions of the variables. The results of the first stage binary regression for determinants of access to credit and access to extension are presented in table 6-2 and 6-3 respectively.

To ensure identification in the estimation of the potential endogenous equation specifications (i.e determinants of access to extension and credit), some variables included in the first-stage estimation are excluded from the second stage equation (i.e hazard model). As suggested by Jacoby and Mansuri (2008), a suitable identification strategy is to employ a variable that strongly influences the endogenous variable (in the access to extension and credit equations), but not the outcome equation (i.e harzard model). In these specifications, extension distance is employed as identifying instrument for access to extension, while distance to farm is employed as identifying instrument for access to credit.

Table 6-2: Maximum likelihood estimates of determinants of access to credit

	Coefficient	S.E
Constant	1.451***	0.374
Baseline hazard	0.006***	0.001
Age	-0.081***	0.007
Farm size	0.404***	0.042
Education	0.0007	0.012
Gender	-0.251*	0.133
Market distance	0.015	0.015
Neighbors	0.656***	0.103
Membership of FBG	0.309***	0.089
Access to extension	-0.058	0.086
Previous adopters in village	-0.009***	0.002
Expected change in the number of adopters	0.035***	0.004
Technology price	0.003***	0.001
Expected change in price	-0.081***	0.280
1st moment of profit	-0.010	0.024
2nd moment of profit	-0.002	0.003
3rd moment of profit	0.354	0.375
4th moment of profit	0.001	0.004
Distance to farm	-0.036*	0.020
Obafemi-Owode	0.136	0.112
Ewekoro	-0.077	0.127
Ifelodun-Irepodun	0.038***	0.130
Loglikelihood	-1830.20	
Likelihood ratio test for model specification	$\chi^2 (21) = 478.22$*** $p\text{-}value$ (0.000)	
Pseudo R^2	0.1156	

*, **, and *** = 10%, 5% and 1% level of significance respectively

Table 6-3: Maximum likelihood estimates of determinants of access to extension

	Coefficient	S.E
Constant	-1.702***	0.443
Baseline hazard	0.003***	0.001
Age	-0.035***	0.005
Farm size	0.370***	0.040
Education	0.002	0.010
Gender	0.238**	0.118
Market distance	-0.023*	0.012
Neighbors	-0.064	0.094
Membership of FBG	0.161**	0.076
Access to credit	-0.108	0.087
Previous adopters in village	0.011***	0.002
Expected change in the number of adopters	0.017***	0.003
Technology price	0.004***	0.001
Expected change in price	-0.156	0.254
1st moment of profit	0.017	0.023
2nd moment of profit	-0.0005	0.003
3rd moment of profit	0.391	0.035
4th moment of profit	0.001	0.004
Extension distance	-0.068***	0.011
Obafemi-Owode	0.700***	0.108
Ewekoro	-0.151	0.131
Ifelodun-Irepodun	-0.147	0.113
Loglikelihood	-2323.976	
Likelihood ratio test for model specification	$\chi^2 (21) = 470.15***$ *p-value* (0.000)	
Pseudo R^2	0.0919	

*, **, and *** = 10%, 5% and 1% level of significance respectively

Extension distance influences access to extension such that farm households which are farther away for the extension office are less likely to receive extension services which in turn determine technology adoption decisions. Thus, extension distance does not have direct relationship with adoption decisions but via extension contact. In the same manner, resource constrained farmers are likely to have their farmsteads farther away from home because land may not be available for them in the nearby villages. The validity of the exclusion restrictions are tested by employing Likelihood ratio test which is carried out by comparing an alternative version of the hazard model that includes the instruments with the one without the instruments. The results of the likelihood ratio test for over identification are reported in table 6-5.

Table 6-4 below shows the descriptive statistics of the sampled farmers. The table highlights arithmetic means as well as differences between adopters and non-adopters of the technology. The *t*-values suggest that these differences are significantly different from zero. Average age of rice farmer is found to be about 44 years with non-adopters (45 years) being a little bit older than adopters (43 years). Difference in age between adopters and non-adopters is significant at 5 percent. The average farm size is about 1.58 hectares, which is consistent with average farm size of smallholder farmers across Africa. However, adopters tend to have larger farm holdings than non-adopters. While adopters cultivated about 1.74 hectares, non-adopters cultivated 1.56 hectares on the average. Difference in farm-size between adopters and non-adopters is significant at 1 percent. Equally, about 70 percent of the adopters indicated that they had fertile soil, whereas only 50 percent of the non-adopters had fertile soil.

Difference in soil quality between adopters and non-adopters is significant at 1 percent. Non-adopters are farther away from markets and extension office than adopters suggesting that they probably have less access to extension information and markets than adopters. The data also show that while 26 percent of the adopters had access to credit, only 19 percent of the adopters had access to credit however, the difference is only significant at 10 percent.

Table 6-4: Descriptive statistics and Definition of variables used in adoption and diffusion models

	Whole sample Mean	Adopters Mean	Non-adopters Mean	Differences
Age of household head (Years)	44.189 (8.247)	43.323 (6.337)	45.021 (9.678)	-1.70**
Farm size (Hectares)	1.577 (1.021)	1.735 (1.092)	1.425 (0.926)	0.310***
Education (Years)	9.045 (4.143)	9.79 (3.638)	8.33 (4.469)	1.460***
Gender (1 if farmer is male, 0 otherwise)	0.889 (0.314)	0.962 (0.191)	0.82 (0.386)	0.143***
Access to credit (1 if farmer is credit unconstrained, 0 otherwise)	0.266 (0.442)	3.118 (0.464)	0.189 (0.416)	0.09*
Market distance (Km)	2.609 (3.205)	2.124 (3.392)	3.075 (2.949)	-0.952***
Membership of farmers based group (1 if farmer is member of FBG, 0 otherwise)	0.479 (0.500)	0.575 (0.496)	0.387 (0.488)	0.187***
Neighbors (1 if farmer interacted with neighbours, 0 otherwise)	0.226 (0.419)	0.274 (0.447)	0.180 (0.386)	0.094**
Extension visits (1 if farmer received extension visits, 0 otherwise)	0.453 (0.498)	0.554 (0.498)	0.356 (0.480)	0.198***
Soil quality (1 if soil is fertile, 0 otherwise)	0.589 (0.492)	0.688 (0.464)	0.494 (0.501)	0.193***
Yield (tons/Ha)	1.846 (0.507)	2.07 (0.037)	1.632 (0.029)	0.437***
Profit (Gross income minus variable costs and depreciation, in Naira ₦)	57461.1 (54679.1)	66919.1 (62072.2)	48393.1 (44822.5)	18526.0***
Location fixed effects				
Obafemi-Owode	0.304 (0.460)	0.323 (0.469)	0.284 (0.452)	
Ewekoro	0.181 (0.386)	0.161 (0.369)	0.201 (0.402)	
Ifelodun-Irepodun	0.179 (0.384)	0.188 (0.392)	0.170 (0.377)	
Gbonyi	0.337 (0.473)	0.328 (0.471)	0.345 (0.477)	

*, **, and *** = 10%, 5% and 1% level of significance respectively.
Exchange rate in 2012, $1 = 160 Naira

89 percent of the rice farmers are men. Although women help their husbands in rice farming and processing, rice is predominantly cultivated by men in the study due to drudgery of the production activities, which may be highly stressful for women. Adopters also obtained higher yields and profits which are statistically significant than those of their non-adopting counterparts; while adopters realize 2.070 tons per hectare, non-adopters realize 1.846 tons per hectare. Similarly, while adopters realized an average profit of ₦66919.10, non-adopters got ₦48393.10 respectively. Differences in yield and profits between adopters and non-adopters are significant at 1 and 5 percent levels respectively. Also statistically significant are the differences in membership of FBG, interaction with neighbors and access to extension between adopters and non-adopters. Adopters and non-adopters also differ in the number of years spent schooling. While the former had approximately 9 years, the latter had 8 years of schooling. Difference in education between adopters and non-adopters is significant at 1 percent.

6.1.2 Empirical results

The non-parametric Kaplan-Meier estimator of survival function was employed to examine subjects' survival function and adoption spells. The approach makes no assumptions regarding the underlying distribution of survival times, it involves measuring the length of time it takes a subject to survive in non-adoption state. A major advantage of the approach is that it takes into account the censoring nature of the data. Estimates show that survival of the sampled farmers decreased with time, though slowly at the beginning, but steeply later. The Log-rank test statistics was used to verify the null hypothesis that membership of FBG and interaction with neighbors does not significantly affect survival function.

The Log-rank test statistics is constructed by computing the observed and expected number of events in one of the groups at each observed event time and then adding these to obtain an overall summary across all time points where there is an event. The hypothesis is rejected at 1 percent level of significance. Kaplan-Meier survivor estimates and Log-rank test statistics results are reported figures 6-2 to 6-5.

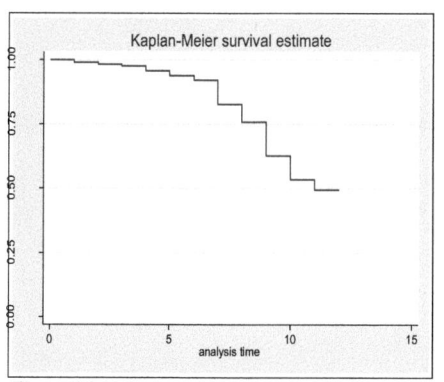

Figure 6-2: Estimates of survival function

Figure 6-3: Effects of FBG on survival function

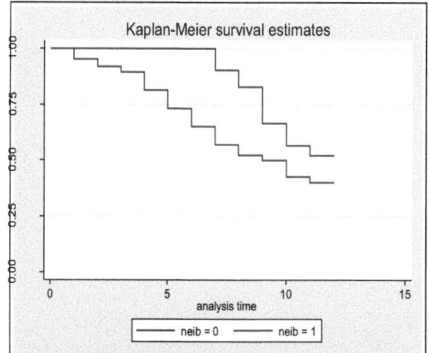

Figure 6-4: Effects of interaction with neighbors on survival function

Log-rank tests for equality of survivor functions			
		Observed	Expected
1	Neighbors=0	1189	1234.71
	Neighbors=1	298	252.29
	Total	1487	1487.00
	Chi2(1)	11.05	Prob>chi2 = 0.0009
2	PVS =0	796	1009.05
	PVS =1	691	477.95
	Total	1487	1487.00
	Chi2(1)	155.10	Prob>chi2 = 0.0000

Figure 6-5: Log-rank tests for equality of survivor functions

The first step taken in the estimation of the hazard function was to choose between discrete-time proportional hazard and logistic duration models for the

99

analysis. The Akaike information criteria (AIC) and Bayesian information criteria (BIC) were employed to ascertain the appropriateness of discrete-time proportional hazard model. The AIC and BIC for proportional hazard model are 861.28 and 1016.89 respectively, while those of the logistic model are 870.746 and 1026.36, respectively, confirming the appropriateness of the proportional hazard model. The maximum likelihood estimates of the parameters of the discrete-time proportional hazard model for technology adoption and corresponding *t*-values are reported in table 6-5. In order to examine the robustness of our estimates, we also estimated a restricted model without the variables summarizing extension services and learning for social networks. In the table, model 1 is full model and model 2 restricted. However, the Likelihood ratio test, as well as both AIC and BIC indicate that the full model is more appropriate for explaining variability in farmers' adoption spells.

Furthermore, Likelihood ratio test was employed to test the null hypothesis that there is existence of unobserved heterogeneity in the model specification. The hypothesis is rejected at the 1% level of significance. The two variables representing the residuals derived from the first-stage regression for the potentially endogenous variables that include access to extension and access to credit are not statistically significant at the conventional levels, indicating no simultaneity bias and that the coefficients have been consistently estimated (Wooldridge, 2010).

The coefficients of the regression can be interpreted as the effects of the covariates on hazard rates or probability of adoption. In this framework, a positive coefficient means that the variable speeds up the adoption process. The coefficient of duration dependence of the baseline hazard is positive and significant, showing that adoption increases with time. The coefficient of age of farmer is negative and significant, implying that younger farmers are more likely to adopt faster than older farmers, a finding that is consistent with earlier findings of Fuglie and Kascak (2001) for the United States and Burton et.al, (2003) for the United Kingdom.

As indicated previously, older farmers tend to be more risk averse, as they may not want to lose yields and profits by abandoning practices and methods that are well known to them. The coefficient of farm size is positive and significant, implying that farmers with bigger farm-holdings have a higher probability of adoption. This is consistent with the findings of Abdulai and Huffman (2005) however, Kashenas and Stoneman (1993) and Genius et al. (2014) found that the coefficient on size is not statistically significant. Gender is positive and significant showing that men are more likely to adopt the technology faster than women. This is probably due to the fact that women in the study area have less access to extension information, since rice production is dominated by men, and extension tends to target male producers.

Age of farmer, farm size, education and gender are jointly significant at the 1% level, showing evidence of rank effects. The variable market distance is negative and significant; suggesting that proximity to market increases the probability of adopting faster, possibly because longer distance to market increases transaction costs and as a result may discourage farmers from market participation. Access to credit is positive and statistically significant, suggesting that farmers that are liquidity constrained are less likely to adopt the technology. If farmers are credit constrained, they tend to devote proportionately less resources to the new technology, especially if the innovation is perceived to be riskier than the traditional technology (Feder and Umali, 1993).

The positive and significant coefficient of previous adopters at village-level shows that learning from other farmers who adopted the technology earlier increases the probability of adoption. The coefficient of price of the technology is not statistically significant at conventional levels. This may be attributed to the price policy of the African Rice Initiative project, which was aimed at promoting NERICA dissemination through provision of certified seeds at subsidized prices. However, the coefficient of expected price is positive and significantly different from zero, indicating that adoption increases as price of technology declines.

The coefficient of the variable representing the expected change in the cumulative number of adopters, used to capture order effects has a positive and statistically significant effect on the probability of adoption. This implies that there are higher and lower order adopters in the producers' population. This may be due to varied access to production information, markets and possession of managerial skills. The empirical results also reveal the significance of access to extension, membership of FBG and interaction with neighbors as these variables are all positive and statistically significant at the conventional levels. Farmers belonging FBG are more likely to adopt the technology than their counterparts not belonging to the group.

Similarly, interaction with neighbors promotes social learning and exchange of views among peers and consequently has positive impact on hazard of adoption. Risk preferences appear to play a significant role in the adoption decisions of the farmers. The first and second moment of profits are positive and statistically significant, suggesting that the higher the expected profit and profit variance, the greater the likelihood of adoption. However, the fourth moment of profits is negative and significantly different from zero, suggesting that farmers consider downside risk which may arise from profit loss and as a results, NERICA technology would only be adopted if found it to have higher profit that the traditional varieties (Koundouri et al. 2006; Genius et al. 2014).

Table 6-5: Maximum-likelihood estimates of the parameters of the hazard model

Variables	Model 1 Coefficients	SE	Model 2 Coefficients	SE
Constant	-13.734***	1.134	-9.544***	0.885
Baseline Hazard	0.054***	0.003	0.043***	0.003
Age	-0.034**	0.015	-0.032**	0.014
Farm size	0.184**	0.088	0.044	0.076
Education	0.056*	0.032	0.113***	0.029
Gender	2.070***	0.493	0.825**	0.405
Market distance	-0.067**	0.028	-0.056**	0.026
Credit	0.064***	0.024	0.061**	0.024
Membership of FBG	0.323*	0.175	-	-
Access to extension	0.756***	0.230	-	-
Neighbors	1.392***	0.211	-	-
Previous adopters in the village	0.048***	0.004	-	-
Expected change in number of adopters	0.108***	0.008	0.096***	0.007
Technology price	0.002	0.002	-0.0005	0.002
Expected change in price	2.788***	0.427	1.899***	0.362
1st moment of profit	0.117***	0.042	0.075*	0.041
2nd moment of profit	0.015***	0.005	0.009*	0.005
3rd moment of profit	0.054	0.661	0.988	0.644
4th moment of profit	-0.018***	0.006	-0.011*	0.006
Credit residuals	-0.123	0.082	-0.130	0.081
Extension residuals	-0.124	0.090	-	-
Location dummies				
Obafemi-Owode	0.959***	0.225	1.122**	0.211
Ewekoro	0.422	0.294	0.412	0.261
Ifelodun-Irepodun	-1.004***	0.279	-0.379	0.234
Number of observations	3731		3731	
Number of respondents	380		380	
Aikake Information Criteria	861.276		1018.072	
Bayesian Information Criteria	1016.88		1142.56	
Loglikelihood	-405.638		-489.036	
Likelihood ratio test for model specification (Wald)	$\chi^2 (26) = 331.90$*** p-value (0.000)			
Likelihood ration test for overidentification	$\chi^2 (2) = 0.310$ p-value (0.858)			

*, **, and *** = 10%, 5% and 1% level of significance respectively

Similarly, interaction with neighbors promotes social learning and exchange of views among peers and consequently has positive impact on hazard of adoption. Risk preferences appear to play a significant role in the adoption decisions of the farmers. The first and second moment of profits are positive and statistically significant, suggesting that the higher the expected profit and profit variance, the greater the likelihood of adoption. However, the fourth moment of profits is negative and significantly different from zero, suggesting that farmers consider downside risk which may arise from profit loss and as a results, NERICA technology would only be adopted if found it to have higher profit that the traditional varieties (Koundouri et al. 2006; Genius et al. 2014).

Table 6-6: Maximum likelihood estimates of hazard ratio

Variables	Hazard ratios	SE	Percentage changes
Baseline Hazard	1.056	0.004	5.60
Age	0.966	0.015	- 3.40
Farm size	1.207	0.106	20.70
Education	1.061	0.034	6.10
FBG	1.377	0.243	37.70
Gender	7.800	3.856	680.00
Market distance	0.937	0.026	- 6.30
Neighbors	4.061	0.859	306.10
Access to extension	2.110	0.486	111.00
Credit	1.067	0.026	6.70
Previous adopter in village	1.049	0.005	4.90
Expected change in adopters	1.115	0.009	11.50
Technology price	1.003	0.002	0.30
Expected change in price	16.446	7.118	1554.60
1st moment of profit	1.120	0.048	12.00
2nd moment of profit	1.015	0.005	1.50
3rd moment of profit	1.004	0.671	0.40
4th moment of profit	0.882	0.072	-11.80

The coefficients in table 6-5 can be converted to hazard ratios by exponentiation to provide a convenient interpretation of these values. Table 6-6 presents hazard ratios of the variables in the discrete time proportional hazard model. A hazard ratio greater than one indicates that an increase in the covariate will lead to an increase in the hazard rate, while hazard ratios below one will cause the hazard rate to decrease. The hazard rate is the conditional probability of adopting the technology in the next short interval given that it has not yet been adopted up to that point. For a specific example, the reported hazard ratio for the variable denoting access to extension (2.110) indicates that farmers who have access to extension services have a conditional probability of adoption which is almost twice that of their counterparts without access to extension.

6.1.3 Concluding remarks

This section investigated impact of rank, stock, order and epidemic effects as well as other farm and non-farm factors on the duration waited by smallholder rice producers to adopt NERICA varieties in Nigeria. The study employed optimal time framework and duration analysis, which bridges the gap between adoption and diffusion studies. Weibull proportional hazard discrete-time duration model was found suitable and used for analysis. Empirical results showed evidence of rank, stock, order and epidemic effects. Rank effect implies that heterogeneity of the rice producers motivated different returns expectations, and as a result, rice producers will embrace the technology at different dates. Order effect suggests that there is existence of adoption orders which might be as a result of varying level of access to information, skills and resources.

Stock and epidemic effect indicates that contact with farmers who had successfully adopted the technology had been a viable medium of communicating the technology to potential adopters. The contributions of membership of FBG and farmer-neighbors interaction as means of promoting learning and exchange of knowledge were also well established. Specifically, differences in farmers' characteristics such as farm size, human capital, age, access to credit, gender and distance from markets were found to be significant in the adoption process. Farmers having larger farm size adopted earlier than those with smaller farm holdings.

Similarly, farmers with more human capital adopted earlier, confirming the widely held view that additional schooling enhances faster adoption. Furthermore, the results showed that farmers' risk preferences tend to influence their adoption behavior, indicating that policy makers need to factor in farmers' risk attitudes when introducing new technologies. Farmers closer to markets adopted faster, a finding that lends further support to the notion that farmers with superior access to input and output markets are normally in a better position to overcome production and marketing constraints that are significant in the adoption of new technologies. Farmers facing liquidity constraints were also found to be less likely to adopt the technology, compared to those that had access to credit to overcome financial constraints and to purchase the required inputs.

The findings from this study suggest that technology dissemination programs should take into consideration the heterogeneity of the farmers that are being targeted with the technology, and as such accompany new technologies with complementary packages like improved access to credits and extension services. To the extent that the participation in group membership where the technology has been introduced and discussed helps to speed up adoption, policy makers could encourage the formation of farmers' groups to enhance the diffusion of new technologies. Improved accesses to output markets, particularly by smallholders, could be facilitated to encourage farmers adopt new technologies.

6.2 Impact evaluation

Impact of NERICA adoption are examined on three outcome variables namely; net returns, poverty head-count and poverty gap using endogenous switching regression approach. Net-returns is calculated as revenue minus variable input costs. As indicated earlier, poverty head-count measures the proportion of farm households living below the poverty line, while poverty gap measures the distance of individual household from the poverty line. Summary statistics of the variables used in econometric analysis is presented in table 6-7, while the full information maximum likelihood estimates of the endogenous switching regression models for joint determinants of adoption and the impact of adoption on the outcome variables are presented in tables 6-8 through 6-10.

6.2.1 Summary statistics and definition of the variables included in the model

The average age of heads of adopting households is found to be 43.32 years, while that of non-adopting households is found to be 45.02 years. The expectation of age of household head on the outcome variables is ambiguous because empirical studies have shown varied findings. For instance, El-Osta and Morehart (2008) showed that age has a linear relationship with poverty incidence amongst farm households in the US, while Cuddy and Paulos (2008) found similar results for smallholder farm households in China. On the contrary, Bogale et al, (2005) found an inverse relationship. These findings can be explained in the context of societal and country level differences. There is tendency for poverty incidence to be high amongst the aged because most of them might have retired from active and high income earning jobs. Conversely, poverty incidence can be high amongst the youths especially if they are not gainfully employed. Therefore, a new variable "age-squared" is created to be able to discern whether there is linear relationship between age of household head and net-returns as well as poverty incidence or not.

Education of the household head is measured by number of years of schooling. According to the human-capital theory, education is expected to have a positive relationship with net-returns and also alleviate poverty because education can enhance allocative ability and efficiency of farmers to critically evaluate characteristics, benefits and costs of technological innovations (Abdulai and Huffman, 2005). Households having many members tend to have a larger labor endowment and are more likely to adopt a new technology and consequently have higher net-returns from rice farming. However, incidence of poverty can be higher in larger households (Moser and Barrett 2003).

A number of studies have shown that probability of technology adoption increases with farm size due to scale effects, as fixed and sunk costs may be lowered by enterprise diversification and adoption of a new technology thereby making the economic unit to operate on a higher profit frontier (see Feder and Umali, 1993). Moreover, farm size and wealth may be positively related because only financially unconstrained farm households can operate a big farm holding. Consequently, larger farm size is hypothesized to have positive impact on net-returns and promote poverty reduction.

As indicated previously, liquidity constraint is measured by a dummy variable indicting if farmers were able to obtain credit when needed. Access to credit facilities enables a farm household to purchase farm inputs and adopt a new technology. Hence, access to credit is expected to have a positive impact on net-returns and negative impact on incidence of poverty. Off-farm income is the income earned from non-farm activities and this is expected to reduce financial constraints, particularly by enabling farmers to purchase productivity-enhancing inputs.

Table 6-7: Summary statistics of farm and household characteristics of adopters and non-adopters of NERICA technology

Variables	Adopters		Non-adopters		Differences
	Mean	SD	Mean	SD	
Outcome variables					
Net-return (Naira)	326434.8	22201.72	236063.9	15697.89	90370.93***
Poverty head-count	0.559	0.036	0.716	0.032	- 0.157***
Poverty gap	0.255	0.021	0.374	0.024	- 0.118***
Explanatory variables					
Age (years)	43.323	6.337	45.021	9.678	-1.70**
Farm size (hectares)	1.735	1.092	1.425	0.926	0.310***
Education (years)	9.790	3.638	8.330	4.469	1.460***
Gender (dummy)	0.962	0.191	0.820	0.386	0.143***
Access to credit (dummy)	3.118	0.464	0.189	0.416	0.09*
Market Distance (Km)	2.124	3.392	3.075	2.949	-0.952***
Membership of farmers' based group (FBG) (dummy)	0.468	0.500	0.284	0.452	0.184***
Neighbors (dummy)	0.274	0.447	0.180	0.386	0.094**
Access to eextension services (dummy)	0.554	0.498	0.356	0.480	0.198***
Soil quality (dummy)	0.688	0.464	0.494	0.501	0. 193***
Output price (Naira)	211.830	2.758	177.445	2.976	34.384***
Household size (number)	5.349	0.168	5.144	0.119	0.205
Off-farm income (dummy)	0.495	0.037	0.299	0.030	0.196***
Land ownership (dummy)	0.575	0.036	0.387	0.035	0.189***
Ownership of livestock (dummy)	0.392	0.035	0.336	0.034	0.057
Fertilizer application (dummy)	0.441	0.037	0.149	0.026	0.291***
Group membership (dummy)	0.608	0.036	0.412	0.036	0.195***
Ownership of radio (dummy)	0.930	0.044	0.629	0.036	0.301***
Localities (districts) dummies					
Obafemi-Owode	0.323	0.469	0.284	0.452	
Ewekoro	0.161	0.369	0.201	0.402	
Ifelodun-Irepodun	0.188	0.392	0.170	0.377	
Gbonyi	0.328	0.471	0.345	0.477	

*, **, and *** = 10%, 5% and 1% level of significance respectively
Exchange rate in 2012, $1 = 160 Naira

Other explanatory variables included in the model are; access to agricultural extension services, social networking, livestock ownership and market distance. Access to agricultural extension services is expected to positively influence adoption decisions, have positive impact on net-returns and contribute to poverty alleviation because agricultural extension is the main source of information to farm households on farming technologies and best agricultural practices. Access and interaction with social networks such as belonging to farmers' cooperatives as well as learning from peers and lead farmers are also expected to have positive impact on technology adoption, net-returns and reduce reduction amongst farm households. Market distance variable is a proxy for market access and extent of transaction costs. Market access is expected to have positive influence on net-returns as farmers are able to sell their produce for rewarding prices and at desirable times. Table 6-7 shows summary statistics of farm and household level characteristics of adopters and non-adopters of NERICA technology.

6.2.2 Determinants of NERICA adoption

The estimates of the determinants of technology adoption are reported in the selection equation columns of tables 6-8 through 6-10. These estimates are the first stage Probit regression, although the results in the three tables are slightly different due to different specifications in the models, the empirical results in the three models are more or less the same and are therefore interpreted together. The results of the selection equations can be interpreted as standard binary probability model. Thus, the positive and significant coefficients of gender, age, farm size, output price, credit, household size, off-farm income, land ownership and fertilizer application indicate that these variables increase probability of adopting NERICA technology.

The positive and significant coefficient of gender can be explained by the fact that men have unrestricted access to social networks and extension services which are the main sources of information about the new technology. Besides, rice is considered as male crop in the study area due to the drudgery involved during the production process. The positive and significant coefficient of age can be explained by the fact that as farmers grow old, they tend to gain increased access to resources (e.g land from inheritance) that aid adoption of technology. However, the negative and significant coefficient of age-squared shows that the relationship between age and probability of adoption is non-linear, rather there exists an evidence of life cycle effects among the farmers suggesting that though innovativeness may grow with age, there comes a point when decline begins to set in. Specifically, the results suggest that the probability of adoption's maximum effect occurs around 46 years.

As indicated previously, farm households with bigger farm size are able to take advantage of economies of scale in production due to benefits such as lower operating costs accruing from expansion and enterprise diversification. They therefore have a higher probability of adoption. Rise or fall in producer price is perceived by farmers as incentive or disincentive for adoption, as such, a higher market price of NERICA encouraged technology adoption.

In the traditional rice farming system, rice production, processing and marketing tasks are usually divided among members of the households. Male household members are involved in tillage operations, female household members normally participate in rice harvesting and processing, while the younger members of the household stay on farm all day to drive pests (birds) away from the rice field, right from the period of fruiting through maturity in order to prevent birds from feeding on immature rice seeds. As a result, large household size is characteristically an important source of labour in rice production enterprise and it is an encouraging factor in adopting the high yielding variety. In the same vein, land ownership and access to production inputs like fertilizer also increase the probability of NERICA technology adoption.

The location fixed effects variables are jointly significant in explaining technology adoption. The results indicate that households in Obafemi-Owode district are more likely to adopt NERICA. A plausible explanation for this is the closeness of most of the communities in the district to the extension office.

6.2.3 Impact of NERICA adoption on net-returns

The empirical results of the differential impact of the explanatory variables on net-returns are presented in the two other columns in table 6-8a. The coefficient of age of household head of the non-adopting households is negative and significant implying that as age increases, net-return decreases. This is consistent with the general notion that productivity decreases as age of farmer increases since smallholder agriculture in most developing countries are usually performed manually (Elias et al, 2013). Gender has a positive and significant impact on net-returns of the adopting households whereas it has no statistically significant influence on net-returns of non-adopting households. This can be attributed to the fact that NERICA is mainly adopted by male headed households.

In both specifications of the outcome equations, number of years of education has positive and significant impact on net-returns. As noted by Penda (2012), education plays a vital role in human capital development as educated folks have a higher tendency to be more productive and are able to take advantage of technological innovations due to their better entrepreneurial and management skills.

The positive and significant coefficients of farm size and output price for both adopting and non-adopting households indicate that as these variables increase, net-returns in rice farming increase accordingly. Farther distance away from produce markets has negative effects on net-returns for both adopting and non-adopting households. This can be explained by increasing transaction and marketing costs. Empirical results also show that off-farm income has positive impact on net-returns in both specifications of the outcome equations. These findings suggest that participation in nonfarm activities tends to enhance the purchasing power of farm households, as income accruing from these activities can be used to purchase productivity-enhancing inputs like labor and fertilizer (Abdulai and Huffman, 2014).

The results also reveal that location fixed effects may be significant in explaining differences in net-returns. In particular, farmers located in Ewekoro tend to have lower net returns, while those located in Obafemi-Owode and Ifelodun-Irepodun are found to have higher net returns. Proper identification of the endogenous switching regression model requires that at least one variable that affect selection but not outcome must be included in the model. Therefore, agricultural extension services, group membership and membership of FBG are used as instruments in this regard. The likelihood ratio test for independent equations is reported in the last row of the table 6-8a. The test's results indicate that the model performed excellently well in explaining determinants of technology adoption, and the differential impact of the explanatory variables on net-returns of both the adopting and non-adopting households.

Finally, at least one of the covariance terms (ρ_A and ρ_{NA}) is statistically significant, indicating that self-selection occurred in technology adoption. Thus, technology adoption may not have the same effect on the non-adopters if they choose to adopt (Lokshin and Sajaia 2004). Moreover, the negative signs of the covariance terms indicate positive selection bias, suggesting that farmers with above-average net-returns have a higher probability of adopting the technology. The results also imply that comparative advantage plays a critical role in the determination of adoption decisions and net returns amongst farm households.

Table 6-8a: Endogenous Switching Regression Results for Determinants of NERICA Adoption and Impact on Net-Returns

Variables	Selection Equation Coefficient	SE	Adopting households Coefficient	SE	Non-adopting households Coefficient	SE
Constant	-5.267***	1.755	7.096**	2.785	-7.976***	1.496
Gender	0.688**	0.323	3.848**	1.594	1.39	0.883
Age	0.275***	0.078	-0.376	0.402	-0.242**	0.117
Age-Squared	-0.003***	0.001	0.004	0.005	-0.004	0.002
Education	0.029	0.117	0.223**	0.107	0.173**	0.096
Farm size	0.215**	0.088	2.985***	0.985	10.292***	1.108
Output price	0.014***	0.002	0.104***	0.018	0.031***	0.01
Credit	0.197***	0.021	0.279	0.703	0.467	0.811
Household size	0.128***	0.045	0.124	0.155	0.202	0.208
Soil quality	0.133	0.171	0.068	0.658	0.263	0.689
Market distance	-0.012**	0.005	-0.216**	0.098	-0.336***	0.115
Off-farm Income	0.457***	0.167	0.544**	0.22	0.573***	0.203
Land ownership	0.284**	0.139	0.332	0.614	0.906	0.714
Ownership of livestock	0.17	0.168	0.098	0.628	0.587	0.678
Fertilizer application	0.938***	0.2	0.179**	0.072	0.026	1.039
Obafemi-Owode	0.275**	0.139	0.176	0.306	0.137*	0.073
Ewekoro	-0.137	0.264	-0.282***	0.151	0.216	0.139
Ifelodun-Irepodun	0.117	0.241	0.210	0.214	0.301**	0.147
Group Membership	0.406**	0.172				
Access to Extension	0.318*	0.163				
Membership of FBG	0.436**	0.169				
$\ln\delta_A$			0.1269	0.021***		
ρ_A			-0.303	0.106***		
$\ln\delta_{NA}$					0.1740	0.024***
ρ_{NA}					0.278	0.164
LR test of independent equations	23.193***	Prob (χ^2) = 0.000				

*, **, and *** = 10%, 5% and 1% level of significance respectively

The mean effect of adoption is reported table 6-8b. Technology adoption significantly increased net-returns of the adopting households as shown by the ATT value of 10.368 compared to 9.648 for non-adopting households representing about 7.5% increase in net-returns of the adopting households. However, the largest impact of the technology adoption on net-returns occurred amongst farmers who cultivated between 0.501 to 1.00 hectares. This implies that farm households cultivating medium sized farms are more productive than their counterparts cultivating smaller or bigger farm holdings.

Table 6-8b: Impact of NERICA adoption on Poverty gap (ATT)

Mean outcomes (log-value)	Adopters	Non-Adopters	ATT	t-value
Net-return	10.368(0.124)	9.648(0.170)	0.721***	4.1708
ATT by Farm Size			Difference	
Less than 0.5 Hectares (38)	1.870(0.646)	0.593(1.771)	1.277**	2.229
≥0.501 – 1.00 Hectares (112)	1.629(0.315)	0.341(0.203)	1.288***	3.448
≥1.01 – 2.00 Hectares (161)	1.406(0.268)	0.623(0.300)	0.783*	1.684
Greater than 2 Hectares (69)	0.804(0.239)	0.605(0.246)	0.1991	0.569

*, **, and *** = 10%, 5% and 1% level of significance respectively
Standard errors are reported in parenthesis

6.2.4 Impact of NERICA poverty head-count

The empirical results of impact of technology adoption on poverty headcount are presented in table 6-9. In this specification, a negative coefficient implies that the variable reduces poverty incidence amongst farm households. Thus, the positive and significant coefficient of gender of household head of non-adopting households suggests that cultivation of traditional rice varieties tends to reduce incidence of poverty amongst women headed households. This can be explained by the fact that NERICA is mainly adopted by men while women still sticks to the cultivation of the traditional varieties because male farmers had unrestricted access to information on the new technology (for example, more extension visits) than their female counterparts due to women marginalization in the rural areas of Nigeria.

The negative and significant coefficient of age of household head of non-adopting households suggests that as the age of the household head increases, poverty incidence tends to reduce amongst non-adopting households, whereas age has no impact on poverty incidence of the adopting households. This suggests that even though traditional rice cultivation has lower economic returns, it tends to reduce poverty incidence amongst older farmers because they have gained mastery of rice farming through many years of farming experience. The coefficient of education of household head of adopting households is negative and significant, suggesting that human capital development plays a significant role in poverty alleviation.

The coefficient of farm size is negative and significant in both specifications of the outcome equations for adopting and non-adopting households alike indicating that households with larger farmer size are able to escape poverty, considering the fact that farm size and wealth are positively related. Higher producer prices and access to credit facilities exhibit poverty alleviating tendencies amongst the adopting households, while the two variables have no impact on non-adopting households. Nevertheless the coefficient of household size in both specifications is positive pointing to the fact that larger farm households tend to be poor. Finally, ownership of livestock has a positive impact on poverty reduction amongst adopting households, but has no impact on non-adopting households.

Proper identification of the endogenous switching regression model requires that at least one variable that affect selection but not outcome must be included in the model. Therefore, membership of farmers' association, access to extension and ownership of radio are used as instruments in this regard. Just as noted above, at least one of the covariance terms (ρ_A and ρ_{NA}) is statistically significant, indicating that self-selection occurred in technology adoption.

Table 6-9a: Endogenous Switching Regression Results for Determinants of NERICA Adoption and Impact on Head-Count

	Selection Equation		Adopting households		Non- adopting households	
Variables	Coefficient	SE	Coefficient	SE	Coefficient	SE
Constant	-5.823***	1.943	0.474***	0.154	1.725***	0.41
Gender	0.707**	0.33	-0.196	0.136	0.118*	0.068
Age	0.292***	0.087	-0.054	0.034	-0.147*	0.086
Age-Squared	-0.003***	0.001	-0.011	0.04	0.017	0.121
Education	0.014	0.027	-0.013**	0.006	-0.006	0.086
Farm size	0.233**	0.116	-0.405***	0.087	-0.647***	0.0827
Output price	0.014***	0.002	-0.005***	0.001	-0.013	0.008
Credit	0.209***	0.097	-0.128**	0.063	-0.077	0.36
Household size	0.026	0.045	0.108***	0.014	0.073***	0.016
Soil quality	0.17	0.168	-0.017	0.058	-0.038	0.051
Market distance	-0.106***	0.026	0.009	0.167	0.021**	0.009
Off-farm Income	0.396**	0.168	-0.124**	0.053	-0.075	0.053
Land ownership	0.161	0.183	-0.015	0.054	-0.024	0.084
Ownership of livestock	0.053	0.16	-0.331	0.252	-0.041	0.051
Fertilizer application	0.215	0.199	0.054	0.069	-0.16	0.063
Obafemi-Owode	0.216**	0.11	-0.079	0.08	-0.133**	0.07
Ewekoro	0.029	0.264	-0.095	0.057	-0.157	0.054
Ifelodun-Irepodun	-0.205	0.223	-0.161**	0.07	-0.042	0.065
Group Membership	0.284*	0.169				
Access to Extension	0.294*	0.151				
Ownership of Radio	0.491***	0.165				
$\ln\delta_A$			0.3422***	0.016		
ρ_A			-0.189	0.341		
$\ln\delta_{NA}$					0.327	0.029***
ρ_{NA}					0.665	0.167***
LR test of independent equations	15.201***	Prob ($\chi2$) = 0.000				

*, **, and *** = 10%, 5% and 1% level of significance respectively

The results also imply that comparative advantage plays a critical role in the determination of net returns from rice farming, poverty incidence amongst farm households as well as adoption decisions. Technology adoption significantly reduced poverty incidence amongst the adopting households as shown by the ATT values of 0.431 compared to 0.672 for non-adopting households. However, the largest impact of technology adoption on poverty reduction occurred amongst farmers who cultivated between 0.501 to 1.00 hectares. As indicated above, farm households cultivating medium sized farms tend to be more productive, as a result, incidence of poverty is lower amongst them.

Table 6-9b: Impact of NERICA Adoption on Head-count (ATT)

Mean outcomes	Adopters	Non-Adopters	ATT	t-value
Head-count	0.431(0.022)	0.672(0.016)	-0.241***	17.141
ATT by Farm Size			Difference	
Less than 0.5 Hectares (38)	-0.291(0.047)	-0.234(0.015)	0.056	1.187
≥0.501 – 1.00 Hectares (112)	-0.282(0.024)	-0.224(0.017)	0.058*	1.872
≥1.01 – 2.00 Hectares (161)	-0.261(0.018)	-0.213(0.021)	0.048*	1.697
Greater than 2 Hectares (69)	-0.247(0.015)	0.212(0.034)	0.035	0.957

*, **, and *** = 10%, 5% and 1% level of significance respectively
Standard errors are reported in parenthesis

6.2.5 Impact of NERICA on Poverty Gap

The empirical results of impact of technology adoption on poverty gap are presented in table 6-10. Just as mentioned above, a negative coefficient implies that variable reduces poverty gap. The negative and significant coefficient of gender of adopting households suggests that cultivation of NERICA rice reduces poverty gap amongst male headed households. This can be explained by the fact that the technology is mostly adopted by male headed households than female headed households. The coefficient of education of household head of adopting households is negative and significant suggesting that human capital development plays significant role in reducing poverty gap amongst adopting households. As indicated previously, educated farmers tend to be more productive and this may increase their farm income and consequently reduce poverty gap.

Table 6-10a: Endogenous Switching Regression Results for Determinants of NERICA Adoption and Impact on Poverty Gap.

Variables	Selection Equation		adopting households		Non-adopting households	
	Coefficient	SE	Coefficient	SE	Coefficient	SE
Constant	-0.765***	1.835	0.0103	0.482	1.169***	0.235
Gender	0.557*	0.319	-0.206**	0.088	0.014	0.038
Age	0.296***	0.084	0.045**	0.022	-0.004	0.009
Age-Squared	-0.003***	0.001	-0.0004**	0.0002	-0.001	0.001
Education	0.012	0.028	-0.013**	0.006	-0.0129	0.039
Farm size	0.233**	0.116	-0.317***	0.056	-0.677***	0.047
Output price	0.014***	0.002	-0.003***	0.0005	-0.017***	0.0001
Credit	0.209**	0.097	-0.061	0.04	-0.051	0.034
Household size	0.049	0.045	0.047***	0.009	0.062***	0.0089
Soil quality	0.174	0.167	-0.004	0.037	-0.026	0.03
Market distance	-0.106***	0.026	0.003	0.006	0.013**	0.005
Off-farm Income	0.396**	0.169	-0.033	0.112	-0.029	0.213
Land ownership	0.161	0.113	-0.026	0.035	-0.002	0.03
Ownership of livestock	0.068	0.163	-0.421	0.302	-0.031	0.029
Fertilizer application	0.103	0.292	0.031	0.041	-0.022	0.046
Obafemi-Owode	0.216**	0.11	-0.02	0.052	0.042	0.044
Ewekoro	-0.057	0.256	-0.071	0.056	-0.111	0.045
Ifelodun-Irepodun	-0.296	0.216	-0.062	0.045	0.011	0.037
Access to Extension	0.294*	0.159				
Membership of FBOs	0.508***	0.152				
$\ln\delta_A$			0.1823***	0.012		
ρ_A			-0.365	0.262		
$\ln\delta_{NA}$					0.231	0.018***
ρ_{NA}					0.789	0.103***
LR test of independent equations	10.228***	Prob (χ2) = 0.0013				

*, **, and *** = 10%, 5% and 1% level of significance respectively

The coefficients of household size of both adopting and non-adopting households are positive and significant suggesting that households with large number of members tend to be poor. This is because as the number of household members increases, per-capital consumption also increases. The coefficients of farm size as well as producer price are negative and significant in both specifications of the outcome equations showing that these variables generally reduce poverty gap amongst the farm households, irrespective of the adoption status. This suggests that higher producer prices increase household income which in turn reduces poverty incidence and gap.

Furthermore, increased household income has a direct link to purchasing power of the farm households. Not only will households with higher purchasing power be able to afford the basic necessities of life, they are also able to invest in their farming operations by adopting yield enhancing technologies. In the same manner, farm households with bigger farm holding are able to produce marketable surplus which tend to increase their household income and close the poverty gap. Proper identification of the endogenous switching regression model requires that at least one variable that affect selection but not outcome must be included in the model. Therefore, agricultural extension services and membership of FBG are used as instruments in this regard.

Table 6-10b: Impact of NERICA adoption on Poverty gap (ATT)

Mean outcomes	Adopters	Non-Adopters	ATT	t-value
Poverty gap	0.141(0.013)	0.309(0.013)	-0.250***	24.759
ATT by Farm Size			Difference	
Less than 0.5 Hectares (38)	-0.433(0.029)	-0.230(0.010)	0.203***	4.234
≥0.501 – 1.00 Hectares (112)	-0.309(0.014)	-0.225(0.012)	0.084***	3.886
≥1.01 – 2.00 Hectares (161)	-0.316(0.017)	-0.160(0.014)	0.156***	8.306
Greater than 2 Hectares (69)	-0.262(0.033)	-0.025(0.010)	0.015	0.591

*, **, and *** = 10%, 5% and 1% level of significance respectively
Standard errors are reported in parenthesis

The results also imply that comparative advantage plays a critical role in the determination of net returns from rice farming, poverty incidence amongst farm households as well as adoption decisions. Technology adoption significantly reduced poverty gap amongst the adopting households as shown by the ATT value of 0.141 compared to 0.309 values of the non-adopting households.

6.2.6 Concluding remarks

Several studies have shown the importance of agricultural growth in poverty reduction in the developing nations of the world. Although most of these studies were carried out in Asian and Latin American countries, there are indications that agricultural growth can also reduce the incidence of poverty in the SSA, where about half of the population lives in abject poverty. However, agricultural growth cannot occur without the adoption of yield enhancing technologies. This section used farm-level data to examine factors that influenced the adoption of NERICA technology, as well as impact of adoption on net returns and incidence of poverty among rice producing households in Nigeria.

Comparisons of farm and household level characteristics between adopting and non-adopting households showed some significant differences. However, these differences are not sufficient to explain the adoption behaviour of rice producing households because the problem of self-selection bias has not been taken into consideration. Endogenous switching regression approach, which accounts for selection bias on both the observables and non-observables characteristics of the farm households in the sampled population, was therefore employed to account for selectivity bias and to estimate the differential impact of technology adoption on the outcome variables of interest.

The empirical results showed that there is endogenous switching and as such, technology adoption may not have the same effect on non-adopters if they chose to adopt the technology. Findings from the study also revealed that formal and extension education played a significant role in technology adoption, net returns and poverty reduction amongst rural households. This suggests that capacity building and human capital development are important for agricultural productivity and poverty reduction. The empirical results also confirmed the importance of social networks and access to production inputs, such as land and fertilizer in adoption decisions and consequently on net returns and poverty incidence of rice producing households.

Availability of labour and credit facilities also exhibit positive and significant relationship with respect to technology adoption. They are therefore key factors in explaining technology adoption, net-returns and poverty alleviation amongst farm households. The results of the causal effect of technology adoption revealed that the adoption of NERICA technology increased net-returns by about 7.5% and reduced poverty by about 35% suggesting that NERICA technology contributed significantly to farm-level productivity, farm income and poverty reduction amongst rice producing households in Nigeria.

6.3 Market participation

Impact of market participation on welfare of rice producing households are examined on three outcome variables namely; return on investment (ROI), poverty head-count and poverty gap, using endogenous switching regression approach. ROI is calculated as profit deflated by investment. In the present study, participating households are defined as farm households that were able to sell their farm produce in various marketing channels in addition to farm gate. On the other hand, non-participants refer to households who are either producing at the subsistent level (i.e, did not sell their farm produce at all) or sold their produce at farm gate.

Summary statistics of the variables used in the econometric analysis is presented in table 6-11, while the full information maximum likelihood estimates of the endogenous switching regression models for determinants of market participation and its impact on the outcome variables are presented in tables 6-12 through 6-14.

6.3.1 Summary statistics and definition of the variables included in the model

Average age of household head of participating household is found to be 43.69 years while that of non-participating households is found to be 47.21 years. Education of the household head is measured by number of years of schooling. While household heads of the participating households have an average of approximately 9.5 years of schooling, heads of non-participating households have about 7 years of schooling. Education is hypothesized to have a positive relationship with ROI and consequently reduce poverty incidence amongst farm households. Education influences the ability to process information and causes farmers to have better access to understanding and interpretation of information (Larpar et al., 2008).

The average farm size of the participating households is found to be 1.42 hectares while that of non-participating households is 1.20 hectares. Farm households with larger farm holdings may participate in market because they are able to produce marketed surplus. Farm households are liquidity unconstrained if they are able to obtain credit if and when needed. While income generated by farm households from non-farm activities is captured as off-farm income. About 32% of the participating households are liquidity unconstrained, while only about 27% of the non-participating households are liquidity unconstrained. Similarly, 52% of the participating households earned off-farm income, while about 30% of the non-participating households earned off-farm income. Access to credit facilities as well as off-farm income can enhance farm household market participation.

Agricultural market information enables farmers to act and make well-informed decisions on where and when to sell their farm produce. Reliable market information can also help farmers in planning production to meet market demand and negotiate rewarding prices. Often times, farmer get market information from their social networks and agricultural extension agents. To this extent, about 57% of the participating households belong to famers' association, while only about 41% of the non-participating household belong to famers' association. Likewise, about 49% of the participating households have access to agricultural extension services, while only about 33% of the non-participating households have access to agricultural extension services.

Ownership of radio and mobile phone can also be variable sources of market information. While about 85% of the participating households own radio set, only about 52% of the non-participating own radio set. Similarly, about 76% of the participating households own mobile phones, while only about 52% of the non-participating households own mobile phones. Market information is hypothesized to have positive impact on market participation and consequently increase ROI and reduce incidence of poverty amongst farm households.

Table 6-11: Summary Statistics of farm and household characteristics of participants and non-participants.

Variables	Participants		Non-Participants		Differences
	Mean	SD	Mean	SD	
Outcome variables					
ROI (Naira)	38163	2884.87	21473	1134.01	16690**
Poverty head-count	0.772	0.028	0.876	0.026	0.104***
Poverty gap	0.420	0.023	0.543	0.027	0.123***
Explanatory variables					
Age (years)	44.690	0.444	47.204	0.829	2.515***
Farm size (Hectares)	1.416	0.093	1.203	0.041	0.213***
Education (years)	9.406	0.232	6.969	0.361	2.437***
Gender (dummy)	0.922	0.018	0.820	0.030	0.102***
Access to credit (dummy)	0.320	0.032	0.267	0.035	0.053
Market Distance (Km)	2.249	0.219	3.839	0.285	1.589***
Access to extension services (dummy)	0.493	0.034	0.329	0.037	0.164***
Soil quality (dummy)	0.626	0.033	0.410	0.039	0.216***
Output price (Naira)	203.571	2.694	179.795	2.694	23.777***
Transportation and processing cost (Naira)	6939.3	640.0	3930.4	520.9	3008***
Household size (number)	5.037	0.134	4.658	0.141	0.380*
Off-farm income (dummy)	0.520	0.034	0.298	0.036	0.222***
Land ownership (dummy)	0.486	0.034	0.354	0.037	0.134**
Group membership (dummy)	0.566	0.033	0.441	0.040	0.125**
Quantity harvested (Kg)	2333.72	87.762	1726.53	102.22	607.19***
Ownership of radio	0.849	0.042	0.516	0.040	0.334***
Ownership of mobile phone (dummy)	0.762	0.027	0.522	0.021	0.240**
Localities (districts) dummies					
Obafemi-Owode	0.352	0.032	0.354	0.037	
Ewekoro	0.155	0.024	0.248	0.034	
Ifelodun-Irepodun	0.178	0.025	0.106	0.024	
Gbonyi	0.315	0.031	0.291	0.036	

*, **, and *** = 10%, 5% and 1% level of significance respectively

6.3.2 Determinants of market participation

The estimates of determinants of market participation are reported in the selection equation columns of tables 6-12 through 6-14. As pointed out previously, the results of the selection equations can be interpreted as standard binary probability model. Thus, the positive and significant coefficients of farm size, output price, quantity harvested and land ownership suggest that the variables increase the probability of market participation. While the negative and significant coefficient of age of household head, market distance and transaction costs show that the variables reduce probability of market participation. The negative and significant coefficient of age can be explained by the fact that as farmers grow old, they tend to have less vigour and resources to carry out marketing activities. This is consistent with the findings of Musara et al. (2011).

The negative and significant coefficient of transaction costs (transportation and processing costs) suggests that the variable has an inverse relationship with the probability of market participation. This is consistent with the findings of Renkow et al., (2013) that farmers tend to opt out of market when profit margins are low due to high transaction costs. Likewise, the negative and significant coefficient of market distance indicates that farm households located far away from the market place are less likely to participate in market.

The positive and significant coefficient of output price shows that farm households have higher probability of market participation when producer prices are higher and better. In the same manner, the positive and significant coefficient of farm size shows that farm households with larger farm holdings are more likely to participate in market because they are able to produce marketable surplus. Access to off-farm income tends to raise farm household's purchasing power thus, the positive and significant coefficient of the variable representing off-farm income suggest that farmers who earned off-farm income have higher probability of market participation.

The positive and significant coefficient of land ownership shows that though the variable may not have direct relationship with market participation, ownership of land and security of tenure tend to encourage longer term farm planning and management which in turns may increase probability of market participation. The location fixed effects variables are jointly significant in explaining market participation. The results indicate that households in Obafemi-Owode are less likely to participate in market and this can be attributed to longer distance of the communities in the district to markets where harvested rice are traded.

6.3.3 Impact of market participation on return on investment (ROI)

The empirical results of the differential impact of the explanatory variables on ROI are presented in the two other columns in table 6-12a. Proper identification of endogenous switching regression model requires that there is at least one variable in the selection or market participation equation that does not appear in the outcome equations. In the ROI specification, the variable representing access to extension and membership of farmers' organization are used as identifying instruments. While access to extension is expected to affect market participation decisions, it should not affect ROI directly. Similarly, membership in a farmer's organization could affect market participation decisions but not ROI.

The coefficient of education of participating households is positive and significant but the variable has no statistically significant influence on ROI of non-participating households. This indicates that human capital development plays a significant role in the ROI of participating households. In both specifications of the outcome equations, the coefficient of household size of both participating and non-participating households is positive and significant suggesting that labour endowments from larger households are capable of enhancing ROI.

Table 6-12a: Endogenous Switching Regression Results for Determinants of Market Participation and Impact on ROI.

Variables	Selection Equation		Participants		Non-Participants	
	Coefficient	SE	Coefficient	SE	Coefficient	SE
Constant	9.0312***	1.324	4.728***	1.365	-13.832	4.866
Age	-0.029**	0.014	-0.124	0.039	-0.049	0.046
Education	-0.027	0.03	0.087**	0.039	-0.095	0.082
Gender	0.125	0.369	1.295	1.124	1.696**	0.824
Credit	0.11	0.239	0.952	0.632	-0.83	0.745
Household size	-0.069	0.548	0.143**	0.059	-0.125**	0.057
Market distance	-0.034***	0.012	-0.083**	0.033	-0.053	0.096
Output price	0.033***	0.004	0.037***	0.012	0.054	0.087
Farm size	0.102***	0.019	1.708***	0.453	1.672***	0.589
Off-farm Income	0.171**	0.084	1.275**	0.53	0.827	0.607
Soil quality	0.057	0.213	-0.458	0.549	0.514	0.59
Transportation and processing cost	-0.076***	0.013	-0.016***	0.005	-0.015	0.004
Land ownership	0.450**	0.207	0.209	0.565	0.054	0.617
Quantity harvested	0.015***	0.003	0.129***	0.016	0.088**	0.034
Obafemi-Owode	-0.736**	0.286	0.433	0.882	2.313***	0.815
Ewekoro	-0.297	0.334	0.62	0.843	3.300***	1.053
Ifelodun-Irepodun	-0.107	0.194	-0.342	0.683	0.956	0.801
Group Membership	0.051***	0.014				
Access to Extension	0.660***	0.194				
$\ln\delta_p$			0.362	0.315		
ρ_p			-0.818***	0.170		
$\ln\delta_{Np}$					0.250	0.214
ρ_{Np}					-0.737***	0.175
LR test of independent equations	18.06***		Prob (χ2) = 0.000			

*, **, and *** = 10%, 5% and 1% level of significance respectively

Market distance variable is a proxy for market access and the possible effects of transaction costs. The empirical results revealed that market distance impacted negatively on ROI of participating households but not on that of non-participants. The results also show that off-farm income has positive impact on ROI of participating households. This suggests that participation in nonfarm activities are variable means of generating incomes which may enhance market participation and consequently increase profitability of rice enterprise. However, the variable does not have a significant impact on ROI of non-participating households. In both specifications of the outcome equations, farm size has positive and significant impact on ROI of both the participating and non-participating households. This implies that farm size has a linear relationship with ROI.

Table 6-12b: Impact of Market Participation on Return of Investment (ROI) (ATT)

Mean outcomes (Logged Value)	Participants	Non-Participants	ATT	t-value
ROI	6.699(0.280)	5.019(0.279)	1.679***	5.9511
ATT by Farm Size			Difference	
Less than 0.5 Hectares (38)	1.6873(0.298)	1.608(0.298)	0.079	0.084
≥0.501 – 1.00 Hectares (112)	1.9374(0.335)	1.0622(0.523)	0.875	1.416
≥1.01 – 2.00 Hectares (161)	1.742(0.372)	1.595(0.435)	0.147	0.396
Greater than 2 Hectares (69)	2.918(0.648)	1.405(0.3117)	1.514**	2.077

*, **, and *** = 10%, 5% and 1% level of significance respectively
Standard errors are reported in parenthesis

Finally, the results show that at least one of the covariance terms (ρ_p and ρ_{Np}) is statistically significant, indicating that self-selection occurred in market participation. Thus, market participation may not have the same effect on non-participants if they choose to participate in markets (Lokshin and Sajaia 2004). Moreover, the negative signs of the covariance terms indicate positive selection bias, suggesting that farmers with above-average ROI have a higher probability of market participation. Thus, comparative advantage tends to play a critical role in the determination of market participation. The mean effects of market participation are reported table 6-12b.

Market participation significantly increased the ROI of the participating households as shown by the ATT value of 6.699 compared to 5.019 value of non-participating households representing about 33.40% increase in the ROI. However, the largest impact of market participation occurred amongst farmers with farm size greater than 2 hectares.

6.3.4 Impact of market participation on poverty head-count

The empirical results of impact of market participation on poverty headcount are presented in table 6-13. In this specification of the outcome equations, a negative coefficient implies that variable reduces poverty. Thus, the negative and significant coefficient of credit for both participating and non-participating households suggests that access to credit can indeed reduce poverty amongst rice producing households.

The empirical results show that large household size may increases poverty incidence. This is confirmed by a positive and significant coefficient of household size in both specifications of the outcome equations. Market distance has a positive and significant impact on poverty head-count of participating households but not on that of non-participants. This shows that the variable possibly increases poverty incidence as longer distance to market tends to discourage market participation due to high transaction cost.

The negative and significant coefficient of output price for participating households signifies that high producer prices reduces poverty amongst participating household but does not have impact on incidence of poverty amongst non-participating households. In both specifications of the outcome equations, the coefficient of quantity of rice harvested for both the participating and non-participating households is negative and significant, suggesting that farmers who produce marketable surplus have a higher tendency to earn income that could help reduce poverty.

Table 6-13a: Endogenous Switching Regression Results for Determinants of Market Participation and Impact on Head-Count.

Variables	Selection Equation Coefficient	SE	Participants Coefficient	SE	Non-Participants Coefficient	SE
Constant	9.237***	1.513	1.357***	0.316	1.365***	0.424
Age	-0.015	0.017	-0.013	0.034	0.003	0.004
Education	-0.027	0.034	-0.007	0.008	-0.046	0.073
Gender	0.403	0.417	-0.019	0.101	0.065	0.072
Credit	0.275	0.268	-0.083**	0.036	-0.109*	0.065
Household size	0.152**	0.063	0.094***	0.012	0.095***	0.014
Market distance	-0.047**	0.022	0.049***	0.013	-0.001	0.009
Output price	0.039***	0.004	-0.003***	0.001	-0.002	0.021
Farm size	-0.012	0.206	-0.08	0.091	-0.277	0.254
Off-farm Income	0.147*	0.083	-0.018	0.047	-0.094	0.055
Soil quality	-0.015	0.026	-0.029	0.049	-0.08	0.052
Transportation and processing cost	-0.084***	0.014	0.014**	0.006	0.025	0.379
Land ownership	0.462	0.336	-0.018	0.151	-0.085	0.055
Quantity harvested	0.177	0.201	-0.105***	0.021	-0.008***	0.002
Obafemi-Owode	-0.584	0.321	-0.051	0.078	-0.131*	0.073
Ewekoro	-0.18	0.372	-0.160**	0.077	-0.201	0.194
Ifelodun-Irepodun	0.186	0.327	-0.063	0.062	0.007	0.069
Group Membership	0.169***	0.026				
Access to Extension	0.508**	0.242				
Mobile phone	0.152***	0.037				
$\ln\delta_p$			0.320***	0.015		
ρ_p			-0.102	0.277		
$\ln\delta_{Np}$					0.308***	0.02
ρ_{Np}					-0.461	0.235
LR test of independent equations	7.37***	Prob ($\chi2$) = 0.007				

*, **, and *** = 10%, 5% and 1% level of significance respectively

Proper identification of the endogenous switching regression model requires that at least one variable that affect selection but not outcome must be included in the model. Therefore, agricultural extension services, group membership and ownership of mobile phones are used as instruments in this regard. Just as noted above, at least one of the covariance terms (ρ_A and ρ_{NA}) is statistically significant, indicating that self-selection occurred in market participation. However, market participation significantly reduced incidence of poverty amongst the participating households as shown by the ATT value of 0.565 compared to 0.658 values of non-participating households representing about 16.46% reduction in poverty as a result of market participation.

Table 6-13b: Impact of Market Participation on Head-count Among Rice Producers

Mean outcomes	Participants	Non-Participants	ATT	t-value
Head-count	0.565(0.234)	0.658(0.017)	-0.093***	8.1629
ATT by Farm Size			Difference	
Less than 0.5 Hectares (38)	-0.121(0.026)	0.117(0.018)	0.238***	6.581
≥0.501 – 1.00 Hectares (112)	-0.023(0.128)	0.148(0.014)	0.164***	6.987
≥1.01 – 2.00 Hectares (161)	0.094(0.011)	0.092(0.018)	-0.002	0.097
Greater than 2 Hectares (69)	0.397(0.026)	0.026(0.009)	-0.371***	16.3283

*, **, and *** = 10%, 5% and 1% level of significance respectively
Standard errors are reported in parenthesis

6.3.5 Impact of market participation on poverty gap

The empirical results of impact of market participation on poverty gap are presented in table 6-14a. Just as mentioned above, a negative coefficient implies that variable reduces poverty gap. Thus, the negative and significant coefficient of education of the household head of participating households suggests the human capital development is important variable in explaining reduction in poverty gap. However, the variable is not significantly different from zero in the case of non-participating households. Access to credit appears to reduce poverty gap of non-participating households while the variable does not have statistically significant impact on poverty gap of participating households.

This is because credit can act as consumption enhancer in addition its use in procuring yield enhancing inputs. In both specifications of the outcome equations, the coefficient of household size of both participating and non-participating households is positive and significant, suggesting that large household size is capable of increasing poverty gap.

Table 6-14a: Endogenous Switching Regression Results for Determinants of Market Participation and Impact on Poverty Gap

Variables	Selection Equation Coefficient	SE	Participants Coefficient	SE	Non-Participants Coefficient	SE
Constant	9.238***	1.438	0.764***	0.209	1.349***	0.462
Age	-1.019***	0.124	0.003	0.002	-0.029	0.275
Education	-0.043	0.036	-0.016***	0.005	-0.002	0.005
Gender	0.393	0.409	-0.033	0.067	-0.065	0.05
Credit	0.295***	0.078	-0.025	0.037	-0.078*	0.046
Household size	0.251***	0.063	0.049***	0.008	0.056***	0.01
Market distance	0.04	0.038	0.005	0.005	0.015	0.062
Output price	0.038	0.004	-0.197***	0.038	-0.026	0.221
Farm size	0.523***	0.103	-0.067**	0.027	-0.249***	0.038
Off-farm Income	1.037***	0.247	-0.052*	0.031	-0.042**	0.018
Soil quality	0.084	0.231	-0.211	0.032	-0.043	0.036
Transportation and processing cost	0.841***	0.139	1.506**	0.681	-0.078	0.382
Land ownership	0.292	0.233	-0.011	0.033	-0.020	0.111
Quantity harvested	1.933***	0.179	-0.058***	0.014	0.099	0.261
Obafemi-Owode	-0.667**	0.33	0.003	0.052	-0.029	0.055
Ewekoro	-0.182	0.362	-0.157***	0.05	-0.173**	0.069
Ifelodun-Irepodun	0.15	0.32	0.018	0.04	0.006	0.049
Access to Extension	0.637**	0.246				
$\ln\delta_p$			0.2118***	0.125		
ρ_p			-0.1253***	0.056		
$\ln\delta_{Np}$					0.2095***	0.01
ρ_{Np}					-0.044	0.2815
LR test of independent equations	12.063***	Prob (χ^2) = 0.000				

*, **, and *** = 10%, 5% and 1% level of significance respectively

The negative and significant coefficient of output price for participating households signifies that high producer prices reduces poverty gap amongst them but the variable does not have effect on incidence of poverty amongst non-participating households. In both specifications of the outcome equations, the coefficient of farm size of both the participating and non-participating households is positive and significant suggesting that the variable is important in explaining reduction in poverty gap amongst them.

The coefficient of transaction cost (transportation and processing costs) of participating households is positive and significant, suggesting that the variable increases poverty gap amongst participating households however, the variable does not have statistically significant impact on incidence of poverty amongst non-participating households.

Market participation significantly reduced poverty gap amongst the participating households as shown by the ATT value of 0.019 compared to 0.177 value of the non-participating households However, the largest impact occurred amongst farmers who cultivated between 1.01 to 2.00 hectares. Proper identification of the endogenous switching regression model requires that at least one variable that affect selection but not outcome must be included in the model. Therefore, access to agricultural extension service is used as instrument in this regard.

Table 14b: Impact of Market Participation on Poverty Gap Among Rice Producers

Mean outcomes	Participants	Non-Participants	ATT	t-value
Poverty Gap	0.019(0.007)	0.177(0.014)	-0.158***	6.9327
ATT by Farm Size			Difference	
Less than 0.5 Hectares (38)	-0.096(0.0078)	-0.010(0.009)	0.085***	5.5858
≥0.501 – 1.00 Hectares (112)	-0.036(0.007)	-0.035(0.011)	0.0014	0.0998
≥1.01 – 2.00 Hectares (161)	-0.142(0.018)	-0.075(0.006)	0.067***	14.376
Greater than 2 Hectares (69)	-0.025(0.011)	-0.008(0.0001)	0.017	1.257

*, **, and *** = 10%, 5% and 1% level of significance respectively
Standard errors are reported in parenthesis

6.3.6 Concluding remarks

The section employed farm-level data to examine factors that influenced market participation by farm households, as well as impact of market participation on ROI and incidence of poverty incidence among rice producing households in Nigeria. Comparison of farm and household level characteristics between households that participated in markets and those that did not participate showed some significant differences. However, these differences are not sufficient to explain the reasons why some households participated in markets while some other ones did not, because the problem of self-selection has not been taken into consideration.

Endogenous switching regression approach which accounts for selection bias on both the observables and non-observables characteristics was therefore employed to account for selectivity bias and to estimate differential impact of market participation on the outcome variables of interest. The results showed that there is endogenous switch and as such, market participation may not have the same effect on non-participants if they chose to participate in market. The empirical results also showed that there is positive selection bias indicating households with above average ROI and wealth were likely to participate in markets. Thus, comparative advantage tends to play a critical role in market participation decisions, ROI and consequently on poverty incidence amongst rice producers.

Empirical results showed that price and non-price factors such as labour, land ownership, access to credit and off-farm income as well as gender and locational characteristics had positive and significant effects in determining market participation, ROI and poverty reduction. Equally, market information variables such as ownership of mobile phone and extension services showed positive and statistically significant impact on market participation. The results of the causal effect of market participation showed that it increased ROI by about 33.47% and reduced poverty by about 16.46% suggesting that market participation contributed significantly to economic returns and poverty reduction of rice producing households in Nigeria.

Chapter Seven

Summary and conclusion

Introduction:

Section 7.1 provides a brief summary and conclusion of the study. Comprehensive reports of the key findings have been provided at the concluding sections of chapter six. Section 7.2 shows the policy implications of this study.

7.1 Summary of findings

The study employed Karshenas and Stoneman (1993) optimal time model and duration analysis to examine farm and non-farm factors affecting adoption and diffusion of New Rice for Africa (NERICA) in Nigeria, while endogenous switching regression approach was employed to analyze impact of NERICA adoption on net-returns from rice farming and incidence of poverty amongst farm households. The study also investigated the determinants and impact of market participation on economic returns and welfare of farm households in order to provide answers to questions on how to facilitate market orientation.

The empirical results showed that adoption and diffusion of NERICA technology were influenced by rank, order, stock and epidemic effects. Production risks and uncertainty also played important roles in farm households' adoption decisions as higher expected profit and profit variance positively and significantly influenced probability of adoption, while access to agricultural extension services, interaction with peers and neighbors were found to be viable media of promoting learning and exchange of knowledge. Access to production inputs and credit also had positive effects on technology adoption.

The results of the causal effects of adoption of NERICA technology on net-returns and incidence of poverty showed that technology adoption increased net returns by about 7.5% and reduced poverty by about 35%, suggesting that adoption of NERICA technology contributed significantly to farm income and welfare of rice producing households in Nigeria. Market participation decisions were influenced by higher producer prices, labour availability, access to credit and off-farm income as well as gender of the household head and locational characteristics. Market information variables such as ownership of mobile phone and access to extension services also had positive and statistically significant impact on market participation. The results of the causal effects of market participation showed that they increased ROI by about 33.47% and reduced poverty by about 16.46% suggesting that market participation contributed significantly to economic returns and poverty reduction among rice producing households in Nigeria.

7.2 Policy implications of the study

The study provided new insights on effective ways of technology dissemination and means of enhancing market participation. The following policy suggestions are crucial to agricultural growth and poverty reduction in Nigeria.

1. It has been established that farm households are endowed with varying levels of resources, they live in different localities and have different production goals. These differences are evident in rank and order effects. This suggests that technology dissemination programs should take into consideration heterogeneity in producers' population. This insight is important in developing strategies for dissemination of new technology.

2. There is evidence of stock and epidemic effects which implies that learning and interaction with peers, and social networking are important media for

technology dissemination. Interaction among farmers and social learning can be encouraged through farmers'-field-days and farmer-field-schools. Since the conventional agricultural extension system in Nigeria is believed to be less effective as far as dissemination of new technologies is concerned, a hybrid system for participatory and interactive learning needs to be developed for a more efficient technology dissemination and capacity building on good agricultural practices.

3. Access to production inputs such as credit and land for farming should be facilitated as this could speed up adoption and diffusion of NERICA technology. Land in particular has been identified as a constraining factor to farming and technology adoption as only farmers with guaranteed access to land have better planning horizon. Therefore, clear policies for agricultural land acquisition by smallholder producers need to be enacted.

4. Transaction cost has been identified as a major constraining factor to market participation. Provision of feeder roads and market infrastructure could ease transportation and movements of produce to market, thereby providing incentives for market participation. As such, much needs to be done on development of rural infrastructure.

5. Finally the role of market information in market participation cannot be over emphasized as this can help in farm operations planning, reduction of wastes and transaction costs. As far as Nigeria is concerned, there are no market information and intelligence services targeted at smallholder producers. Therefore, policy makers could integrate this aspect in the national agricultural extension system.

References

Abadie, A. (2005). Semiparametric Difference-in-Differences Estimators. *The Review of Economic Studies*, 72(1), 1-19.

Abdulai, A. and C. R. Binder (2006). Slash-and-burn cultivation practice and agricultural input demand and output supply. *Environment and Development Economics* 11 (2): 201–20.

Abdulai, A. and E. A. Birachi (2009). Choice of Coordination Mechanism in the Kenyan Fresh Milk Supply Chain. *Applied Economic Perspectives and Policy*, *31*(1), 103-121.

Abdulai, A. and W. Huffman (2005). The Diffusion of New Agricultural Technologies: The Case of Crossbred-Cow Technology in Tanzania. *American Journal of Agricultural Economics*, 87(3), 645-659.

Abdulai, A., and W. Huffman (2014). The Adoption and Impact of Soil and Water Conservation Technology: An Endogenous Switching Regression Application. *Land Economics*, 90(1), 26-43.

Abdulai, A., Monin, P. and J. Gerber (2008). Joint Estimation of Information Acquisition and Adoption of New Technologies under Uncertainty. *Journal of International Development* 20(4), 437-451.

Adekambi, S. A., Diagne, A., Simtowe, F. P., and G. Biaou (2009). The Impact of Agricultural Technology Adoption on Poverty: The Case of NERICA Rice Varieties in Benin. In International Association of Agricultural Economists' 2009 Conference, Beijing, China, August 16 (Vol. 22, p. 2009).

Adeoti J.O. and B.T. Sinh (2009). Technological Constraints and Farmers' Vulnerability in Selected Developing Countries (Nigeria and Vietnam). Paper Presented at the 7th International Conference 2009, 6-8 October, Dakar, Senegal.

Africa Rice Center (2008). The New Rice for Africa - A Compendium. Somado E. A, R. G., Guei and S. O., Keya eds. Africa Rice Center (WARDA), Cotonou; Food and Agriculture Organization of the United Nations, Rome; Sasakawa Africa Association, Tokyo. 210 p.

Akindele, S. T. and A. Adebo (2004). The Political Economy of River Basin and Rural Development Authority in Nigeria: A Retrospective Case Study of Owena-River Basin and Rural Development Authority (ORBRDA). *Journal of Human Ecology*, *16*(1), 55-62.

Akintayo, I., Ajayi, O., Agboh-Noameshie R. A. and B. Cissé (2009). 7th African Rice Initiative Steering Committee/Experts Meeting May 4-6, 2009, Cotonou, Benin, Report, IRAT, 1967. Les variétés de riz au cercle de Banfora. IRAT/Haute-volta. L'"Agronomie Tropical, V. 24, p. 691-707.

Akpokodje, G., Lancon, F.and O. Erenstein (2001). Nigeria's Rice Economy: State of the Art. Paper Presented at the Nigerian Institute for Social and Economic Research (NISER)/West Africa Rice Development Association (WARDA), Nigeria Rice Economy Stakeholders Workshop Ibadan, 8-9 November, 55pp.

Aksoy, M. A. and Beghin, J. C. (2004). Global Agricultural Trade and Developing Countries. World Bank Publications.

Ali, A. and A. Abdulai (2010). The Adoption of Genetically Modified Cotton and Poverty Reduction in Pakistan. *Journal of Agricultural Economics*, 61(1), 175-192.

Allison, P. D. (1982). Discrete-Time Methods for the Analysis of Event Histories. *Sociological methodology*, 13(1), 61-98.

American Marketing Association (AMA) (1985). New Marketing Definition, Marketing News, No. 5, 1 March 1985.

Amsalu, A. and J. De Graaff (2007). Determinants of Adoption and Continued Use of Stone Terraces for Soil and Water Conservation in an Ethiopian Highland Watershed. *Ecological Economics*, 61(2), 294-302.

Antle, J. M. (1983). Testing the Stochastic Structure of Production: A Flexible Moment-Based Approach. *Journal of Business & Economic Statistics*, 1(3), 192-201.

Antle, J. M. (2010). Do Economic Variables Follow Scale or Location-Scale Distributions? *American Journal of Agricultural Economics*, 92(1), 196-204.

Asfaw, S., Mithöfer, D. and H. Waibel (2009). Investment in Compliance with GLOBALGAP Standards: Does It Pay Off for Small-Scale Producers in Kenya?. *Quarterly Journal of International Agriculture*, 48(4), 337-362.

Bacha, D., Namara, R., Bogale, A., and A Tesfaye. (2011). Impact of Small-Scale Irrigation on Household Poverty: Empirical Evidence from the Ambo District in Ethiopia. *Irrigation and Drainage*, 60(1), 1-10.

Baidu-Forson, J. (1999). Factors Influencing Adoption of Land-Enhancing Technology in the Sahel: Lessons from a Case Study in Niger. *Agricultural Economics*, 20(3), 231-239.

Baptista, R. (1999). The Diffusion of Process Innovations: A Selective Review. *International Journal of the Economics of Business*, 6(1), 107-129.

Barrett, C. B. (2008). Smallholder Market Participation: Concepts and Evidence from Eastern and Southern Africa. *Food Policy*, 33(4), 299-317.

Bass, Frank M. (1969).A New Product Growth Model for Consumer Durables. *Management Science*, 15 (5), 215-227.

Becerril J. and A. Abdulai (2010). The Impact of Improved Maize Varieties on Poverty in Mexico: A Propensity Score Matching Approach. *World Development*, 38(7), 1024–1035.

Bellemare, M. F. and C. B. Barrett (2006). An Ordered Tobit Model of Market Participation: Evidence from Kenya and Ethiopia. *American Journal of Agricultural Economics*, 88(2), 324-337.

Bezemer, D. and D. Headey (2008). Agriculture, Development, and Urban Bias. *World Development*, 36(8), 1342-1364.

Blandford, D. (1979). West African Export Marketing Boards. *Agricultural Marketing*.

Blundell, R. and M. Costa Dias (2000). Evaluation Methods for Non-Experimental Data. *Fiscal Studies*, 21(4), 427-468.

Bogale, A., Hagedorn, K. and B. Korf (2005). Determinants of Poverty in Rural Ethiopia. *Quarterly Journal of International Agriculture*, 44(2), 101-120.

Bryson, A., Dorsett, R. and S. Purdon (2002). The Use of Propensity Score Matching in the Evaluation of Active Labour Market Policies, Department for Work and Pensions Working Paper No.4.

Burton, M., Rigby, D., and T. Young (2003). Modeling the Adoption of Organic Horticultural Technology in the UK Using Duration Analysis. *Australian Journal of Agricultural and Resource Economics*, 47, 29–54.

Busch, L. (1988). Universities for Development: Report of the Joint Indo-U.S. Impact Evaluation of the Indian Agricultural Universities. A.I.D. USAID - Project Impact Evaluation Report No. 68. Washington, DC.

Byerlee, D., de Janvry, A. and E. Sadoulet (2009). Agriculture for Development: Toward a New Paradigm. *Annual Review of Resource Economics*, 1(1), 15-35.

Byerlee, D. and E. H. de Polanco (1986). Farmers' Stepwise Adoption of Technological Packages: Evidence from the Mexican Altiplano. American *Journal of Agricultural Economics*, 68(3), 519-527.

Bzugu, P. M., Mustapha, S. B., and E. A. Zubairu (2010). Adoption of NERICA 1 Rice Variety among Farmers in Jalingo Local Government Area of Taraba State, Nigeria. *Journal of Environmental Issues and Agriculture in Developing Countries*, 2 (2), 132-139.

Cadoni, P. and F. Angelucci (2013). Analysis of Incentives and Disincentives for Rice in Nigeria. Technical Notes Series, MAFAP-FAO, Rome.

Caliendo, M and S. Kopeinig (2008). Some Practical Guidance for the Implementation of Propensity Score Matching. *Journal of Economic Surveys*, 22. (1), 31 - 72.

Cameron, L. A. (1999). The Importance of Learning in the Adoption of High-Yielding Variety Seeds. *American Journal of Agricultural Economics,* 81(1), 83-94.

Central Intelligence Agency (CIA), (2014). The World Fact-Book 2014. Available at *https://www.cia.gov/library/publications/the-world-factbook/geos/ni.html. Accessed on 02/08/2014*.

Chang, H. H. (2013). Old Farmer Pension Program and Farm Succession: Evidence from a Population-Based Survey of Farm Households in Taiwan. *American Journal of Agricultural Economics*, 95(4), 976-991.

Chukwuemeka, E. and H. N. Nzewi (2011). An Empirical Study of World Bank Agricultural Development Programme in Nigeria. *American Journal of Social and Management Sciences*, 2(1), 176-187.

Conley, T. G. and C. R. Udry (2001). Social Learning through Networks: The Adoption of New Agricultural Technologies in Ghana. *American Journal of Agricultural Economics*, 668-673.

Conley, T. G. and C. R. Udry (2010). Learning about a New Technology: Pineapple in Ghana. *The American Economic Review*, 35-69.

Coudouel, A., Hentschel, J. S. and Q. T. Wodon (2002). Poverty measurement and analysis. *A Sourcebook for poverty reduction strategies*, 1, 27-74.

Cox, D.R. (1972). Regression Models and Life Tables. *Journal of Royal Statistical Society*, 34:187–220.

D'Emden F.H., Llewellyn R.S., and M.P. Burton (2008). Factors Influencing Adoption of Conservation Tillage in Australian Cropping Regions. *Australian Journal of Agriculture and Resource Economics*, 52 (2), 169–182.

Dadi, L., Burton, M. and A. Ozanne (2004). Duration Analysis of Technological Adoption in Ethiopian Agriculture. *Journal of Agricultural Economics*, 55 (3), 613-631.

Daramola, A., Ehui, S., Ukeje E. and J. McIntire (2007). Agricultural Export Potential in Nigeria. Economic Policy Options for a Prosperous Nigeria. Palgrave: Macmillan. Available at *http://www.csae.ox.ac.uk/books/APfP-series/epopn/AgriculturalexportpotentialinNigeria.pdf*

De Janvry, A. and E. Sadoulet (2002). World Poverty and the Role of Agricultural Technology: Direct and Indirect Effects. *Journal of Development Studies*, 38 (4), 1–26.

De Janvry, A. (2010). Agriculture for Development: New Paradigm and Options for Success. *Agricultural Economics*, *41*(s1), 17-36.

Deaton, A. (2010). Instruments, Randomization, and Learning about Development. *Journal of Economic Literature*, 424-455.

Diagne, Aliou (2006). The Diffusion and Adoption of NERICA Rice Varieties in Cote d'Ivoire. *The Developing Economies,* 44(2) 208-231.

Dibba, L., Diagne, A., Fialor, S. C. and F. Nimoh (2012). Diffusion and Adoption of New Rice Varieties for Africa (NERICA) in the Gambia. *African Crop Science Journal*, 20(1), 141 – 153.

Di Falco, S. and J. P. Chavas (2009). On Crop Biodiversity, Risk Exposure, and Food Security in the Highlands of Ethiopia. *American Journal of Agricultural Economics*, *91*(3), 599-611.

Dorward, A., Kydd, J. and C. Poulton (2005). Institutions, Markets and Economic Development: Linking Development Policy to Theory and Praxis. *Development and Change*, 36(1), 1–25.

Doss, C. R. (2006). Analyzing Technology Adoption Using Micro-Studies: Limitations, Challenges, and Opportunities for Improvement. *Agricultural Economics,* 34(3), 207-219.

Duflo, E. and M. Kremer (2005). Use of Randomization in the Evaluation of Development Effectiveness. *Evaluating Development Effectiveness*, *7,* 205-231.

Eggleston, K., Sun, A. and Z. Zhan (2014). The Impact of Rural Pensions in China on Migration and Off-farm Employment of Adult Children and Extended Households' Living Arrangements. Working Papers, Walter H. Shorenstein Asia-Pacific Research Center, Stanford University.

El-Osta, H. S., and M. J. Morehart (2008). Determinants of Poverty among US Farm Households. *Journal of Agricultural and Applied Economics*, 40(1), 1-20.

Encyclopedia Britannica Online. Available at *http://www.britannica.com/EBchecked-/topic/414840/Nigeria/55285/Climate.* Accessed 05-08-2014.

Ersado, L., Amacher, G. and J. Alwang (2004). Productivity and Land Enhancing Technologies in Northern Ethiopia: Health, Public Investments, and Sequential Adoption. *American Journal of Agricultural Economics*, 86(2), 321-331.

Etim, N. A. and G. E. Edet (2013). Constraints of the Nigerian Agricultural Sector: A Review. *British Journal of Science*, 10(1), 22 – 32

Fafchamps, M. and R. V. Hill (2005). Selling at the Farm-gate or Traveling to Market. *American Journal of Agricultural Economics*, 87(3), 717-734.

FAOSTAT (2013). Online Database of the Statistics Division of Food and Agriculture Organization of the United Nations *http://faostat.fao.org/faostat/* (Accessed on 15 December, 2013)

Feder, G. and D. L. Umali (1993). The Adoption of Agricultural Innovations: A Review. *Technological Forecasting and Social Change*, 43(3), 215-239.

Feder, G., Murgai, R. and J. Quizon (2004). The Acquisition and Diffusion of Knowledge: The Case of Pest Management Training in Farmer Field Schools, Indonesia. *Journal of Agricultural Economics*, 55(2), 217–239.

Feleke, S. and T. Zegeye (2006). Adoption of Improved Maize Varieties in Southern Ethiopia: Factors and Strategy Options. *Food Policy*, 31:442–457.

FGD, (2012). Focus Group Discussion with Rice Producing Households during the Field Survey in Ogun and Ekiti States, Nigeria.

Fischer, E., and M. Qaim (2012). Linking Smallholders to Markets: Determinants and Impacts of Farmer Collective Action in Kenya. *World Development*, 40(6), 1255-1268.

Foster, J., Greer, J. and E. Thorbecke, E. (1984). A Class of Decomposable Poverty Measures. *Econometrica*, 761-766.

Freeman, H. A., Ehui, S. K. and M. A Jabbar (1998). Credit Constraints and Smallholder Dairy Production in the East African Highlands: Application of a Switching Regression Model. *Agricultural Economics*, *19*(1), 33-44.

Genius, M., Koundouri, P., Nauges, C. and V. Tzouvelekas (2014). Information Transmission in Irrigation Technology Adoption and Diffusion: Social Learning, Extension Services, and Spatial Effects. *American Journal of Agricultural Economics* 96(1), 328–344.

Goetz, S. J. (1992). A Selectivity Model of Household Food Marketing Behaviour in Sub-Saharan Africa. *American Journal of Agricultural Economics*, 74(2), 444-452.

Gourlay, A. and E. Pentecost (2002). The Determinants of Technology Diffusion: Evidence from the UK Financial Sector. The Manchester School, 70(2), 185-203.

Gridley, H., Jones, M. P., and M. Wopereis-Pura (2002). Development of New Rice for Africa (NERICA) and Participatory Varietal Selection (PVS). Bouaké, Côte d'Ivoire: West Africa Rice Development Association (WARDA).

Griliches, Z. (1957). Hybrid Corn-An Exploration in the Economics of Technical Change. *Econometrica* 48, 501-522.

Gubert, F., Lassourd, T. and S. Mesplé-Somps (2010). Do Remittances Affect Poverty and Inequality? Evidence from Mali. *Analyse à Partir de Trois Scénarios Contrefactuels. Revue Economique*, 61(6), 1023-1050.

Haggblade, S. and P. Hazell (1989). Agricultural Technology and Farm-Nonfarm Growth Linkages. *Agricultural Economics,* 3(4), 345-364.

Heckman, J. J. (1979). Sample Selection Bias as a Specification Error. *Econometrica*, 153-161.

Heckman, J. and B. Singer (1984). A Method for Minimizing the Impact of Distributional Assumptions in Econometric Models for Duration Data. *Econometrica: Journal of the Econometric Society*, 271-320.

Heckman, J. J. and S. Urzua (2010). Comparing IV with Structural Models: What Simple IV Can and Cannot Identify. *Journal of Econometrics*, 156(1), 27-37.

IFDC, (2008). Study of the Domestic Rice Value Chains in Mali, Niger, and Nigeria, West Africa. International Fertilizer Development Centre. U.S.A.

IFPRI, (2010). Quantitative Analysis of Rural Poverty in Nigeria, Nigeria Strategy Support Program, Brief No. 17. International Food Policy Research Institute, USA.

Igben, M. S. and A. C. Nwosu (1987). Issues and Problems in the Administration of the Ministry Of Agriculture-Based Extension Service in Nigeria. *Agricultural Administration and Extension*, 27(4), 215-230.

Iheme, D.A. (1996). The Marketing of Staple Food Crops in Enugu State, Nigeria: A Case Study of Rice, Maize and Beans. An M.Sc. Thesis Submitted to the faculty of Agriculture, University of Nigeria, Nsukka.

Imbens, G. W. and J. D. Angrist (1994). Identification and Estimation of Local Average Treatment Effects. *Econometrica*, 467-475.

Imbens, G. W. and W. K. Newey (2009). Identification and Estimation of Triangular Simultaneous Equations Models without Additivity. *Econometrica*, 77(5), 1481-1512.

Imbens, G. W. (2009). Better LATE than nothing: Some comments on Deaton (2009) and Heckman and Urzua (2009). *Journal of Economic Literature*, 48(2), 399-423

Jacoby, H. G., and G. Mansuri (2008). Land Tenancy and Non-Contractible Investment in Rural Pakistan. *The Review of Economic Studies*, 75(3), 763-788.

Jayne, T., Mather, D. and E. Mghenyi (2010). Principal Challenges Confronting Smallholder Agriculture in Sub-Saharan Africa. *World Development*, 38(10), 1384-1398.

Jenkins, S.P. (1995). Easy Estimation Methods for Discrete-Time Duration Models. *Oxford Bulletin of Economics and Statistics*, 57(1), 129-136.

Johnson, N. L. and S. Kotz (1970). Distributions in Statistics: Continuous Univariate Distribution Vols. 1 and 2. Boston: Houghton Mifflin.

Jones, M. P. and Wopereis-Pura Myra (Eds.), (2002). Participatory varietal selection. Beyond the Flame. WARDA, Bouaké, Cote d'Ivoire, 80pp.

Kabunga, N. S., Dubois, T., and M. Qaim (2012). Yield Effects of Tissue Culture Bananas in Kenya: Accounting for Selection Bias and the Role of Complementary Inputs. *Journal of Agricultural Economics*, 63(2), 444-64.

Karshenas, M. and P. Stoneman (1993). Rank, Stock, Order and Epidemic Effects in the Diffusion of New Process Technology. *RAND Journal of Economics*, 24(4), 503–528.

Kassie, M., Shiferaw, B. and G. Muricho (2011). Agricultural Technology, Crop Income, and Poverty Alleviation in Uganda. *World Development*, 39(10), 1784-1795.

Kebbeh, M., Haefele, S., Fagade, S. O., and C. D. I. Abidjan (2003). Challenges and Opportunity of Improving Irrigated Rice Productivity in Nigeria, West African Rice Development Association. Abidjan, Cote d'Ivoire. Available at *http://pdf.usaid.gov/pdf_docs/Pnadb849.pdf.* Accessed on 12th July, 2013.

Khandker, S. R., Koolwal, G. B. and H. A. Samad (2010). Handbook on impact evaluation: quantitative methods and practices. World Bank Publications.

Kijima, Y. K., Otsuka, D. and D. Sserunkuuma (2008). Assessing the Impact of NERICA on Income and Poverty in Central and Western Uganda. *Agricultural Economics,* 38 (3), 327–337.

Kleemann, L., Abdulai, A., and M. Buss (2014). Certification and Access to Export Markets: Adoption and Return on Investment of Organic-Certified Pineapple Farming in Ghana. *World Development*, *64*, 79-92.

Koundouri, P., Nauges, C. and V. Tzouvelekas (2006). Technology Adoption under Production Uncertainty: Theory and Application to Irrigation Technology. *American Journal of Agricultural Economics*, *88*(3), 657-670.

Kydd, J. and A. Dorward, (2004). Implications of Market and Coordination Failures for Rural Development in Least Developed Countries. Journal of International Development. 16(7), 951–970.

Lancaster L. (1990). The Econometric Analysis of Transition Data. Cambridge University Press, New York.

Leathers, H. D. and M. Smale (1991). A Bayesian Approach to Explaining Sequential Adoption of Components of a Technological Package. *American Journal of Agricultural Economics*, 73(3), 734-742.

Lee, L. F. (1982). Some Approaches to the Correction of Selectivity Bias. *Review of Economic Studies* 49: 355-72.

Ligon, E. and S. Elisabeth (2008). Estimating the Effects of Aggregate Agricultural Growth on the Distribution of Expenditures. Background paper for the WDR 2008.

Lipsey, R. G and P. O. Steiner (1981). "Economics (6th ed.)". New York: Harper and Row Publishers, 1981.

Lipton, M. (1988). The Place of Agricultural Research in the Development of Sub-Saharan Africa. *World Development,* 16(10), 1231-1257.

Lokshin, M., and Z. Sajaia (2004). Maximum Likelihood Estimation of Endogenous Switching Regression Models. *Stata Journal*, 4, 282-289.

Longtau, Selbut R. (2003). Nigeria Case Study Report on Rice Production. Multi-Agency Partnerships for Technical Change in West African Agriculture (MAPS). Jos, Nigeria. Eco-Systems Development Organization (EDO) for Overseas Development Institute (ODI).

Lubungu, M. (2013). Welfare Effects of Smallholder Farmers' Participation in Livestock Markets In Zambia. In 2013 Annual Meeting, August 4-6, 2013, Washington, DC (No. 150606). Agricultural and Applied Economics Association.

Maddala, G. S. (1986). Limited-Dependent and Qualitative Variables in Econometrics. New York: Cambridge University Press.

Maertens, A. and C.B. Barrett (2013). Measuring Social Networks' Effects on Agricultural Technology Adoption. *American Journal of Agricultural Economics*, 95(2), 353-359.

Mansfield, E. (1961). Technical Change and the Rate of Imitation. *Econometrica: Journal of the Econometric Society*, 741-766.

Marra, M., Pannell, D. J. and A. Abadi Ghadim (2003). The Economics of Risk, Uncertainty and Learning in the Adoption of New Agricultural Technologies: Where Are We on the Learning Curve? *Agricultural Systems*, 75(2), 215-234.

Matsumoto, T., Yamano, T. and D. Sserunkuuma (2013). Technology Adoption and Dissemination in Agriculture: Evidence from Sequential Intervention in Maize Production in Uganda (No. 13-14). National Graduate Institute for Policy Studies.

Mendola, M. (2007). Agricultural Technology Adoption and Poverty Reduction: A Propensity Score Matching Analysis for Rural Bangladesh. *Food Policy*, 32 (3), 372–93.

Minten, B. and C. B. Barrett (2008). Agricultural Technology, Productivity, and Poverty in Madagascar. *World Development*, 36(5), 797-822.

Munshi, K. (2004). Social Learning in a Heterogeneous Population: Social Learning in the Indian Green Revolution. *Journal of Development Economics*, 73, 185-213.

Murage, A. W., Obare, G., Chianu, J., Amudayi D. M., Pickett J. and Z. R. Khan (2011). Duration Analysis of Technology Adoption Effects of Dissemination Pathways: A Case of 'Push-Pull' Technology for Control of Striga Weeds and Stem Borers in Western Kenya. *Crop Protection*, 30 (5), 531-538.

Musara, J. P., Zivenge, E., Chagwiza, G., Chimvuramahwe, J. and P. Dube (2011). Determinants of Smallholder Cotton Contract Farming Participation in a Recovering Economy: Empirical Results from Patchway District, Zimbabwe. *Journal of Sustainable Development in Africa,* 13(4).

National Rice Development Strategy (NRDS) (2009). A Working Document Prepared for the Coalition for African Rice Development. Federal Ministry of Agriculture and Rural Development, Abuja, Nigeria. Available at http://www.inter-reseaux.org/IMG/pdf_NRDS_FINAL__National_rice_development_stategy_.pdf

National Rice Survey (2009). Rice Data System in Nigeria, National Rice survey 2009, Nigeria Bureau of Statistics, Abuja, Nigeria.

NBS (2014). Nigerian Gross Domestic Product Report Quarter One 2014, Nigeria Bureau of Statistics, Abuja, Nigeria.

Nguezet, P. M., Diagne, A., Olusegun Okoruwa, V. and V. Ojehomon (2011). Impact of Improved Rice Technology (NERICA Varieties) on Income and Poverty among Rice Farming Households in Nigeria: A Local Average Treatment Effect (LATE) Approach. *Quarterly Journal of International Agriculture,* 50(3), 267.

Nguezet, P. M., Diagne, A., Olusegun Okoruwa, V. and V. Ojehomon (2013). Estimating the Actual and Potential Adoption Rates and Determinants of NERICA Rice Varieties in Nigeria. *Journal of Crop Improvement,* 27(5), 561-585.

Norris, E. P. and S. S. Batie (1987). Virginia Farmers' Soil Conservation Decisions: An Application of TOBIT Analysis. *Southern Journal of Agricultural Economics,* 19:79-90

Ogun State MANR (2012). NERICA Dissemination Progress Report. Ogun State Agricultural Development Programme (OGADEP), Ibara, Abeokuta, Nigeria.

Ojehomon, V. E. T., Adewumi, M. O., Omotesho, O. A., Ayinde, K. and A. Diagne (2012). Adoption and Economics of New Rice for Africa (NERICA) among Rice Farmers in Ekiti State, Nigeria. *Journal of American Science*, 8(2), 423-429.

Ojiako, I. A., Manyong, V. M. and A. E. Ikpi (2007). Determinants of Rural Farmers' Improved Soybean Adoption Decision in Northern Nigeria. *Journal of Food, Agriculture and Environment*, 5 (2), 215-223.

Okoruwa, V. O. and O. O. Ogundele (2004). Technical Efficiency Differentials in Rice Production Technologies in Nigeria. *World Journal of Agricultural Sciences,* 3(5), 53 – 58.

Okuneye, P. A. and I. A. Idowu (1990). Rural Development Strategies in Nigeria: Past Experiences and Some Lessons. *Journal of Rural Cooperation*, 18(1), 31-54.

Olomola, A. S. (2007). Competitive Commercial Agriculture in Africa Study (CCAA): Nigeria Case Study. Final Report submitted to the Canadian International Development Agency (CIDA) and the World Bank.

Omamo, S. W. (1998). Transport Costs and Small Holder Cropping Choices: An Application to Siaya District, Kenya. *American Journal of Agricultural Economics*, 80(1), 116-123.

Omilola, B. (2009). Estimating the Impact of Agricultural Technology on Poverty Reduction in Rural Nigeria (Vol. 901). International Food Policy Research Institute.

Oni, O., Nkonya, E., Pender, J., Phillips, D. and E. Kato (2009). Trends and drivers of agricultural productivity in Nigeria. International Food Policy Research Institute (IFPRI).

Oruonye, E. D. and E. Okrikata (2010). Sustainable use of plant protection products in Nigeria and challenges. *Journal of Plant Breeding and Crop Science*, 2(9), 267-272.

Oteng, J. W. and Sant'Anna, R. (1999). Rice Production in Africa: Current Situation and Issues. International Rice Commission Newsletter V.48.

Ouma, E., Jagwe, J., Obare, G. A., and S. Abele (2010). Determinants of Smallholder Farmers' Participation in Banana Markets in Central Africa: The Role of Transaction Costs. *Agricultural Economics*, 41(2), 111-122.

Pinstrup-Andersen, P. and P. Hazell (1985). The Impact of the Green Revolution and Prospects for the Future. *Food Reviews International*, 1(1), 1-25.

Ransom, J. K., Paudyal, K. and K. Adhikari (2003). Adoption of Improved Maize Varieties in the Hills of Nepal. *Agricultural Economics* 29(3), 299-305.

Ravallion, M. and S. Chen (2007). China's (Uneven) Progress against Poverty. *Journal of Development Economics*, 82(1), 1-42.

Renkow, M., Hallstrom, D. G. and D. D Karanja (2004). Rural Infrastructure, Transactions Costs and Market Participation in Kenya. *Journal of Development Economics*, 73(1), 349-367.

Rogers, E. M. (2003). Diffusion of Innovations (5th edition.) New York, Free Press.

Rogers, E. M. (1995). Diffusion of innovations (4th editon.). New York: The Free Press.

Romer, P. (1994). New Goods, Old Theory and the Welfare Cost of Trade Restrictions. *Journal of Development Economics*, 43(1), 5-38.

Roseboom, J., Beintema, N. M., Pardey, P. G. and E.O. Oyedipe (1994). The National Agricultural Research System of Nigeria. Statistical Brief No. 15. The Hague: ISNAR.

Rosenbaum, P. and D. B. Rubin, (1983). The Central Role of the Propensity Score in Observational Studies for Causal Effects. *Biometrika*, 70(1), 41-55.

Rossi, Peter H. and Howard E. Freeman (1993). Evaluation: A Systematic Approach, Beverly Hills, California: SAGE Press.

Rossini, A. J. and A. A. Tsiatis (1996). A Semi-Parametric Proportional Odds Regression Model for the Analysis of Current Status Data. *Journal of the American Statistical Association*, 91(434), 713-721.

Rubin, D. B. (1974). Estimating Causal Effects of Treatments in Randomized and Nonrandomized Studies. *Journal of educational Psychology*, 66(5), 688.

Rubin, D. B. (1977). Assignment to Treatment Group on the Basis of a Covariate. *Journal of Educational and Behavioural statistics*, 2(1), 1-26.

Rubin, D.B (2004). Teaching Statistical Inference for Causal Effects in Experiments and Observational Studies. *Journal of Educational and Behavioural Statistics*, 29(3), 343-367.

Rui Baptista, (1999). The Diffusion of Process Innovations: A Selective Review: *International Journal of the Economics of Business*, 6(1), 107-129.

Ruttan, Vernon W. and Hayami Yujiro, (1972). Strategies for Agricultural Development, Food Research Institute Studies, Stanford University, Food Research Institute, Issue 02.

Sarkar, J. (1998). Technological Diffusion: Alternative Theories and Historical Evidence. *Journal of Economic Survey,* 12 (2), 131-176.

Shahidur, R., Gayatri, B. and A. Hussain (2010). Handbook on Impact Evaluation: Quantitative Methods and Practices. The World Bank, Washington DC.

Shakya, P. B. and J. C. Flinn (1985). Adoption of Modern Varieties and Fertilizer Use on Rice in the Eastern Tarai of Nepal. *Journal of Agricultural Economics*, 36(3), 409–419.

Shenyang, G. and M. W. Fraser (2010). Propensity Score Analysis: Statistical Methods and Applications. Advanced Quantitative Techniques in the Social Sciences. SAGE Publications.

Shimada, S. (1999). A Study of Increased Food Production in Nigeria: The Effect of the Structural Adjustment Program on the Local Level. *African Study Monographs*, 20(4), 175-227.

Sidibé, A. (2005). Farm-Level Adoption of Soil and Water Conservation Techniques in Northern Burkina-Faso. *Agricultural Water Management,* 71(3), 211-224.

Smith, R. J. and R. W. Blundell (1986). An Exogeneity Test for a Simultaneous Equation Tobit Model with an Application to Labor Supply. *Econometrica*, 54:679–685.

Smith, J. and P. E Todd (2005). Does Matching Overcome Lalonde's Critique of Nonexperimental Estimators? *Journal of Econometrics*, 125(1), 305-353.

Stoneman, P. L. and P. A David (1986). Adoption Subsidies Vs Information Provision as Instruments of Technology Policy. *The Economic Journal*, 142-150.

Strauss, J. (1984). Marketed Surpluses of Agricultural Households in Sierra Leone. *American Journal of Agricultural Economics*, 66(3), 321-331.

Tiamiyu, S. A., Akintola J. O. and M. A. Y Rahji (2009). Technology Adoption and Productivity Difference among Growers of New Rice for Africa in Savannah Zone of Nigeria. *Tropicultura,* 27(4), 193–197.

UNIDO, CBN, BOI (2010). Unleashing Agricultural Development in Nigeria through Value Chain Financing. Working Paper, November 2010. United Nations Industrial Development Organization (UNIDO). Vienna, Austria.

United Nation, (2013). Economic Report on Africa (2013). United Nations Economic Commission for Africa, Addis Ababa. Available at *http://www.uneca.org/sites/default/files/uploaded-documents/era2013_casestudy_eng_nigeria.pdf.*

Von Oppen, M., Njehia B. and A. Ijaimi (1997). Policy Arena: The Impact Market Access on Agricultural Productivity: Lessons from India, Kenya and Sudan. *Journal of International Development*, 9(1), 177-131.

Witcombe, J. R. (1996). Participatory Approaches to Plant Breeding and Selection. *Biotechnology and Development Monitor*, *29*, 2-6.

Wooldridge, J. M. (2002). Econometric Analysis of Cross Section and Panel Data. Cambridge, MA: MIT Press.

World Bank (2001). Global Economic Prospects and the Developing Countries 2001. Washington, D.C.

World Bank (2013). World Data Bank - Poverty and Inequality Database. Available at *http://databank.worldbank.org/data/views/variableselection/selectvariables.aspx?source=Poverty-and-Inequality-Database* accessed on 14th December, 2013.

World Bank, (2002). World Development Report (2002): Building Institutions for Markets. Oxford University Press, New York.

World Bank, (2005). World Development Indicators 2005. The World Bank, Washington, D.C.

Zhao, Z. (2003). Data Issues of Using Matching Methods to Estimate Treatment Effects: An Illustration with NSW Data Set. China Centre for Economic Research. 0: aucune, 1.

Zilberman, D. (1984). Technological Change, Government Policies, and Exhaustible Resources in Agriculture. *American Journal of Agricultural Economics, 66*(5), 634-640.

I want morebooks!

Buy your books fast and straightforward online - at one of the world's fastest growing online book stores! Environmentally sound due to Print-on-Demand technologies.

Buy your books online at
www.get-morebooks.com

Kaufen Sie Ihre Bücher schnell und unkompliziert online – auf einer der am schnellsten wachsenden Buchhandelsplattformen weltweit!
Dank Print-On-Demand umwelt- und ressourcenschonend produziert.

Bücher schneller online kaufen
www.morebooks.de

OmniScriptum Marketing DEU GmbH
Heinrich-Böcking-Str. 6-8
D - 66121 Saarbrücken
Telefax: +49 681 93 81 567-9

info@omniscriptum.com
www.omniscriptum.com

Printed by Books on Demand GmbH, Norderstedt / Germany